MANUAL OF
Water Fluoridation Practice

FRANZ J. MAIER

Sanitary Engineer, Div. of Dental Public Health and Resources, USPHS
Fellow, American Society of Civil Engineers
Fellow and Life Member, American Public Health Association
Member, American Water Works Association
Diplomate, American Academy of Sanitary Engineers
Member, Asociación Interamericana de Ingeniería Sanitaria
Licensed Professional Engineer, District of Columbia

McGRAW-HILL BOOK COMPANY, INC.
New York   Toronto   London

MANUAL OF WATER FLUORIDATION PRACTICE

Copyright © 1963 by the McGraw-Hill Book Company, Inc. All Rights Reserved. Printed in the United States of America. This book, or parts thereof, may not be reproduced in any form without permission of the publishers. *Library of Congress Catalog Card Number* 62-20724

39718

Preface

Increasingly, groups of people representing many different backgrounds and interests have been brought together by a mutual interest in water fluoridation. They have in common a desire to learn why more than 40 million persons in the United States have adopted fluoridation. Such divergent groups include members of town councils and city manager's offices; medical and dental organizations; engineers in health, waterworks, and public works departments at all levels of government; consulting engineering firms; members of service clubs that have sponsored fluoridation; students of engineering, dentistry, and related disciplines; and individuals concerned about the health and well-being of their fellow citizens.

This book not only provides an explanation for such groups of why fluoridation is desirable but suggests methods for implementing it by providing answers to such specific questions as:

What would be the probable cost, not only of the fluoridation equipment but also the chemical and maintenance costs?

What are the benefits to be expected, and when and where should one look for them?

What arguments will be used to discourage the adoption of fluoridation?

What engineering problems have been encountered elsewhere and how have these been solved?

What steps need be taken to determine the best and most economical installation?

What kind of help might be needed and where can it be found?

The intention of this book, then, is to help anyone in a community learn what he can expect to gain by adopting fluoridation, how he should go about getting it, and what it will cost him.

As an aid in estimating costs, descriptions are given of the many variations in methods for adding fluorides to a water supply, together with the accessory equipment which is necessary or desirable. A chapter is included on how to determine the quantity of fluorides in water both before and after treatment.

For those people living in communities with water supplies which naturally contain too many fluorides, a chapter is included describing how such water is treated. Another chapter is provided describing how families with their own private water supplies can fluoridate their water.

Acknowledgment is of course given to many scientists, particularly those associated with the U.S. Public Health Service, who have contributed much of the basic information in this book.

Franz J. Maier

Contents

Preface iii

1. *History and Growth* 1
   Fluorosis Observations and Research; Discovery of Role of Fluorides; Determination of Optimum Fluoride Concentrations; Inception of First Fluoridation Projects.
2. *Benefits of Fluoridation* 14
3. *Adoption Procedures* 26
4. *Fluoridation Litigation* 32
5. *Technical Objections (Engineering, Chemical, Industrial, Economic)* 39
6. *Fluoride Concentration Desired in the Treated Water Supply* 48
7. *Fluoride Compounds (Characteristics, Sources, Costs)* . . 60
8. *Feeder Types and Capacities* 91
9. *Feeder Auxiliary Equipment* 125
10. *Points of Application* 155
11. *Control of Fluoride Concentration (Laboratory Procedures)* 158
12. *Safety of the Water-plant Operators* 188
13. *Effects of Fluorides on the Distribution System* 195
14. *Present Status of Fluoridation* 200
15. *Fluoridation of Individual Water Supplies* 204
16. *The Practicality of Partial Defluoridation* 209

APPENDIX: *Health Objections* 215

Index 227

CHAPTER 1 *History and Growth*

Over thirty years ago, at several widely scattered places in the world, it was noticed that relatively small groups of people had a significantly lower susceptibility to a particular disease. Such a phenomenon was very unusual, and many investigators tried to discover the cause. The reason for this was later found to be that the water consumed at these places contained a peculiar ingredient. The effectiveness of the ingredient depended on its concentration; too much produced an undesirable result, too little was ineffective. Immediately many people wondered whether adding the ingredient in the right amounts to their drinking water would result in a similar reduction in this illness. This was tried in several places, and it was in fact found to work just as well as if the ingredient had been there naturally. As a result, many places started adding this material to their water supplies. This, in brief, is the history of fluoridation as it relates to the control of dental decay.

Reduction in dental caries, however, was not the first indication of the effects of fluorides in water. Dr. J. M. Eager, a U.S. Public Health Service physician stationed in Venice, had observed that certain Italian emigrants from a region nearby had peculiarly marked teeth. In 1901 he wrote that the black teeth (*denti di Chiaie*) he observed were popularly believed to have been caused by using waters "charged under pressure with volcanic fumes" or by the fumes themselves. He also observed that "strong well-formed teeth not particularly prone to decay appear to be the rule among young Italians when they have not been subjected to the influence during infancy of the causes of 'Chiaie's disease.'"

In its mildest form this disease is characterized by very slight, opaque, whitish areas on some of the posterior teeth. As the defect becomes more severe, the mottling is more widespread and changes in color through the grays to black. In addition, in the most severe cases, gross calcification defects occur, resulting in attrition of the enamel. In some of the latter cases the teeth quickly deteriorate so badly that they wear down to the gum line, and complete dentures must be obtained.

Several observers reported such defects among children in Colorado, Italy, and England. In 1916 Dr. Frederick S. McKay, a practicing dentist, reported that many of his patients in Colorado Springs, Colorado, had this defect. Later, by studying the many degrees of severity of mottling, he arrived at the conclusion that it must be caused by some as yet unreported substance in the drinking water. He was so convinced of this that he recommended that the water supply of Oakley, Idaho, be changed because of the severe mottling occurring among the children there. The supply was changed in 1925 to a nearby spring which had been used by a few other children who were free of mottling. A similar study at Bauxite, Arkansas, resulted in a change of their water supply in 1928. This study revealed the probable role of drinking water by indicating mottling among people reared on water which had been used since 1909, when a change was made in the source of their water; prior to this date, no mottling had occurred. In addition, those whose tooth enamel had developed elsewhere and who had then moved to Bauxite had no mottling.

An interesting sidelight on Dr. McKay's conviction regarding the role of water supplies was his plea to waterworks experts for help in interpreting water analyses.[1] In 1926 he reviewed his accumulated evidence leading to the implication of water supplies and ruled out any connection between the calcium or iron content of water and the incidence of mottling.

Replies solicited from prominent water-treatment experts typified the opinion of most observers as to the probable causes of mottling; that is, theories as to the ingredient in water which might contribute to the cause were founded on manganese, acidity, pH, hardness, organic material, and, in addition, unbalanced diet and the effect of the mother's diet on the dental health of the child. The most sig-

[1] Frederick S. McKay, Water Supplies Charged with Disfiguring Teeth, *Water Works Eng.*, **79**(2): 71 (1926).

nificant reply came from Frank Hannan,[2] the chemist at the Toronto, Ontario, filtration plant. Mr. Hannan said, in part:

> Since the enamel is essentially mineral in composition and the water definitely incriminated, the mineral content of the water seems the probable source of the trouble. Of the mineral elements at present known to be common to both water and enamel, the chief ones are calcium, phosphorus, and fluorine. For our intake of phosphorus, we are independent of the small proportion found in water; the same can be asserted of calcium with perhaps a shade less certainty, a dietary deficient in this element being not altogether unusual. But when we consider fluorine, all is at present shrouded in obscurity. The detection and estimation of small traces of fluorine are tedious and troublesome and quite outside the province of the ordinary water works chemist who has to handle many thousands of samples in a year. The French chemist Gautier found fluorine practically universally present in water. That it cannot exist there in more than traces is fortunately secured to us by the low solubility product of calcium fluoride, fluorine in sizable doses being a rather powerful poison. The U.S. Public Health Service seem indicated to handle a many-sided, time-consuming research of this kind. Should the incriminated waters prove to be all alike fluorine-free, the case for fluorine deficiency will become strong.

Unfortunately, Dr. McKay dismissed Mr. Hannan's suggestion with "None of the elements mentioned by Mr. Hannan seem to have any direct possibility of producing a decalcification."[3]

Collaborating with other investigators, in 1928 Dr. McKay stated:

> My testimony has been supplemented by that of others, who report that these mottled enamel cases, in the various districts, are singularly free of caries. . . . The great majority of cavities consisted of small pit and fissure lesions in the molars and seldom did caries extend beyond that stage. In this respect the behavior of dental caries in the mouths of these children is distinctly different to that which usually occurs.

Several more years had to elapse before the cause of mottled enamel was discovered, almost simultaneously, by three different groups of scientists working independently with entirely different tools and methods—and at such widely separated places as Pittsburgh, Pennsylvania, Arizona, and North Africa.

[2] Frank Hannan, Do Certain Water Supplies Disfigure the Teeth? *Water Works Eng.*, **79**(14): 934 (1926).
[3] Frederick S. McKay, Do Water Supplies Cause Defects in Teeth Enamel? *Water Works Eng.*, **79**(20): 1332 (1926).

A. W. Petrey, a chemist with the Aluminum Company of America, noticed the calcium fluoride band in a spectroscopic examination for aluminum in a water sample from Bauxite, Arkansas. The chief chemist of these laboratories, H. V. Churchill, later reported during 1931 that similar examinations of water samples from areas where mottled enamel was endemic invariably showed the presence of fluorides.

At almost the same time, Drs. H. V. Smith, M. C. Smith, and E. M. Lantz reported the cause of mottling by duplicating the lesion in rats by means of concentrating naturally fluoridated water and comparing the results with the lesions observed when a diet high in fluorides was used. Also during 1931 H. Velu in North Africa showed that mottling could be produced in animals by using waters saturated with natural phosphates from Algiers and Morocco, the lesions being identical to those produced by feeding rats a high-fluoride diet.

Dr. Churchill concluded that endemic regions had waters containing 2 ppm (parts per million) or more fluoride while those areas without mottling had water supplies with less than 1.0 ppm. This division of fluoride waters was confirmed by the Smiths in Arizona, who reported that water sources from nonendemic areas contained less than 0.72 ppm fluoride.

The term "ppm" is a measure of the concentration of a mineral or other ingredient in a liquid, gas, or other solid. One part per million fluoride in water, for example, means that in every million parts by weight of water there is one part by weight of fluoride ion. One part per million fluoride is equivalent to $8\frac{1}{3}$ lb fluoride ion per million gal water because 1 gal water weighs approximately $8\frac{1}{3}$ lb. In metric units, one part per million is identical to one milligram per liter (the weight of a liter of pure water being one kilogram).

The chemical element known as fluorine is a gas that combines actively with other elements to form fluoride compounds. Elemental fluorine is practically never found in nature, but compounds containing fluorides are found almost everywhere. Fluorine constitutes approximately 0.077 per cent of the earth's crust and as such ranks thirteenth among the elements in order of abundance. Sea water contains about 1.4 ppm, which makes fluoride rank twelfth in order of concentration. In the human body only a trace exists, but nevertheless here also it is thirteenth in abundance. The most commonly found fluoride minerals are fluorspar (containing fluorite or calcium fluoride), cryolite (containing the double fluoride salt of sodium and aluminum), and apatite (which is a complex calcium compound of fluorides, carbonates, and sulfates). When water passes over or through deposits of these or similar fluoride-containing compounds, a portion is dissolved and the water then carries a quantity (measured in ppm) of fluorides and other ions.

When acids, bases, or salts are dissolved in water, they are broken up or dissociated into minute particles called ions, which consist of atoms or groups of atoms. Ions are electrically charged whether or not an electric current has been passed through the solution.

If, however, an electric current is passed through the solution containing the dissolved substance, the charged particles are carried by the current and discharged at the electrodes. The negatively charged particles, or anions, are carried to the positively charged pole, or anode; the positively charged particles, or cations, are carried to the negatively charged pole, or cathode. By the discharge of the ions at the electrodes their charges are neutralized electrically. In this manner the particles cease to be ions with an electrical charge and become instead atoms or groups of atoms with no charge at all.

If a substance such as hydrochloric acid (HCl) is dissolved in water, its dissociation may be represented by the equation:

$$HCl \rightleftarrows H^+ + Cl^-$$

In other words, each of the molecules of hydrochloric acid that dissociates forms a positively charged hydrogen ion and a negatively charged chloride ion. The double arrow indicates that this reaction is reversible, or capable of proceeding in either direction. If hydrochloric acid is dissolved in water, a part is dissociated, or the reaction proceeds from the left to the right. On the other hand, if hydrogen ions and chloride ions are brought together in a solution, a certain proportion of them will combine to form hydrochloric acid.

Similarly hydrogen ions, if brought in contact with oxygen ions, will form molecules of water. The hydrogen ions are identical in both cases and can form many hundreds of different compounds in combination with other elements. The same is true of other ions; that is, they are identical in every respect to every other one of the same name. It is for this reason that fluoride ions, no matter what their source, are identical regardless of which compound they were taken from. For example, the ions used for forming sodium fluoride are exactly the same as those in calcium fluoride, ammonium silicofluoride, potassium fluoride, hydrofluosilicic acid, and many hundreds of other fluoride-containing compounds.

Because fluoride-bearing minerals are so widely distributed over the earth, it would be expected that water containing fluorides would also be found almost everywhere. After the discovery of the role of fluorides in water in producing mottling, the interest in fluoride analysis of water markedly increased. As these analyses accumulated, it was found, in fact, that fluoride waters were almost everywhere— from Pikes Peak to Death Valley, from Alaska to South Africa. The waters of all the oceans contain between 1 and 1.5 ppm fluoride.

It is almost impossible for anyone to identify definitely the compound which provides the source of the fluoride ion in a fluoride water. This is because of the tremendous quantities and infinite varieties of fluoride minerals scattered everywhere under and on the earth. Analysis of water reveals only the chemical elements contained in a sample—never the chemical compounds. Whenever an analyst attempts to reassemble into compounds the elements he has determined, the term "hypothetical" is applied to the compounds thus suggested. For instance, if he finds two different positive and two different negative elements, he can say that possibly two compounds were originally present because he can form two hypothetical combinations from the four elements found. Generally, an analysis of a water reveals that innumerable hypothetical combinations are possible.

Most of the fluoride-containing minerals are forms of fluorspar, apatite, or cryolite. It might be assumed then that most of our fluoride waters are derived from their percolation over and through such minerals and dissolving them in their passing.

As more data became available on the large number of water supplies containing fluorides, it became evident that a large number of people had been drinking water containing fluorides and that the severity of mottling varied with the fluoride level in the water.

At this point Dr. McKay[4] could establish the following facts:

1. Mottled enamel can be produced only during the period of calcification of the tooth, not thereafter. In other words, after a certain age (about 12) mottled enamel cannot be produced, whatever the level of fluoride in the water.

[4] Frederick S. McKay, Mottled Enamel: Early History and Its Unique Features, from "Fluorine and Dental Health," American Association for the Advancement of Science, Washington, 1942.

History and Growth

2. Once such lesions are formed, they cannot be repaired, either during the calcification period or thereafter. No medicinal or dietary influence can alleviate the disfigurement.

3. Fluorine appears to be the only agent ordinarily a part of the diet which has an influence on enamel formation.

4. After calcification is completed, the structure of the enamel remains unaltered despite changes in diet.

It was in refining this relationship that Dr. H. Trendley Dean[5] of the U.S. Public Health Service began his carefully prepared and executed epidemiological investigations. His first task was to determine

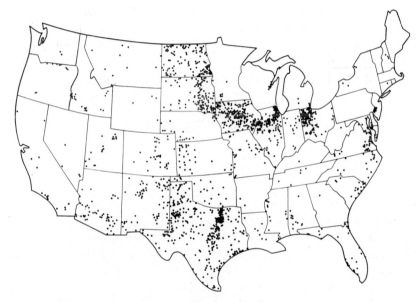

FIG. 1.1. Communities using naturally fluoridated water containing 0.7 ppm or more fluoride. (*USPHS.*)

the extent of mottled enamel (the term was changed to "fluorosis" after its cause became known) and establish the direct relationship between it and the fluoride concentrations of the drinking waters used. He confirmed that many localities have water supplies containing fluorides (Fig. 1.1). Areas with the largest number of such supplies and containing the highest levels of fluorides include those states running from North Dakota to Texas, those along the Mexican border,

[5] H. Trendley Dean, The Investigation of Physiological Effects by the Epidemiological Method, from "Fluorine and Dental Health," American Association for the Advancement of Science, Washington, 1942.

TABLE 1.1. MOTTLED-ENAMEL EXPERIENCE

City	Fluoride	Number examined	Norm. ques.	Very mild	Mild	Moderate	Moderately severe	Severe	Incidence	Reference
Ankeny, Iowa	8.0	21	0	0	2	10	7	2	100.0	3
Britton, S. Dak.	6.6	63	6	11	10	26	−10−		90.0	4
Lake Preston, S. Dak.	5.9	50	2	5	18	19	−6−		96.0	4
Post, Tex.	5.7	38	0	0	4	19	13	2	100.0	3
Hecla, S. Dak.	5.0	14	1	2	3	8	−0−		93.0	4
Pierpont, S. Dak.	4.7	26	3	7	9	7	−0−		88.0	4
Lubbock, Tex.	4.4	176	2	11	54	77	31	1	98.8	2
Conway, S.C.	4.0	59	7	12	19	14	7	0	88.2	3
Amarillo, Tex.	3.9	168	6	30	58	60	12	2	96.4	2
Platte, S. Dak.	3.0	55	19	12	12	10	−2−		65.0	4
Plainview, Tex.	2.9	78	2	24	24	26	2	0	87.6	2
St. Lawrence, S. Dak.	2.9	28	7	6	6	9	−0−		75.0	4
Doland, S. Dak.	2.9	25	9	2	10	4	−0−		64.0	4
Ipswich, S. Dak.	2.8	53	20	15	12	5	−1−		63.0	4
Redfield, S. Dak.	2.7	48	19	14	5	9	−1−		60.0	4
Carthage, S. Dak.	2.6	34	16	8	8	2	−0−		52.0	4
Onida, S. Dak.	2.6	31	20	6	5	0	−0−		35.0	4
Colorado Springs, Colo.	2.5	54	18	14	10	11	1	0	66.6	1
Wolsey, S. Dak.	2.4	16	4	2	6	4	−0−		75.0	4
Gettysburg, S. Dak.	2.4	53	24	16	8	3	−2−		55.0	4
Leola, S. Dak.	2.3	31	15	10	6	0	−0−		51.0	4
Clovis, N. Mex.	2.2	138	40	33	49	15	1	0	71.0	3
Miller, S. Dak.	2.2	51	26	15	7	3	−0−		49.0	4, 6
Iroquois, S. Dak.	2.1	28	10	8	5	4	−1−		64.0	4, 6

Location	F									Ref
Galesburg, Ill. (first study)	1.9	39	24	11	3	1	—	0	38.5	1
Galesburg, Ill. (second study)	1.9	243	129	82	29	—3—		0	46.9	5
Elmhurst, Ill.	1.8	170	102	51	15	2	—0—	0	40.0	7
Monmouth, Ill. (first study)	1.7	29	15	12	2	0		0	48.3	1
Monmouth, Ill. (second study)	1.7	99	32	47	20	—0—		0	67.7	5
Aberdeen, S. Dak.	1.7	166	129	24	10	3	—0—	0	22.0	4,6
Webster City, Iowa	1.6	47	31	13	3	0		0	34.0	3
East Moline, Ill.	1.5	94	70	18	5	1	0	0	25.5	3
Howard, S. Dak.	1.4	48	39	6	3	6	—0—	0	18.0	4,6
Joliet, Ill.	1.3	447	334	99	14	0	—0—	0	25.3	7
Maywood, Ill.	1.2	171	114	50	7	0	—0—	0	33.3	7
Aurora, Ill.	1.2	633	538	88	7	0	—0—	0	15.0	7
Mullins, S.C.	0.9	47	42	4	1	0	0	0	10.6	3
Big Spring, Tex.	0.7	68	66	2	0	0	0	0	2.9	2
Junction City, Kans	0.7	93	91	2	0	0	0	0	2.1	3
Pueblo, Colo.	0.6	49	47	2	0	0	0	0	4.0	1
Elgin, Ill.	0.5	403	386	14	3	0	0	0	4.2	7
Macomb, Ill.	0.2	63	62	1	0	0	0	0	1.6	5
Quincy, Ill.	0.2	291	291	0	0	0	0	0	0.0	5
Lake Michigan—Evanston, Oak Park, Waukegan Ill.	0.0–0.1	1,008	1,001	0	0	0	0	0	0.2–1.6	7

"Highest concentration of fluoride incapable of producing a definite degree of mottled enamel in as much as 10% of the group examined."

"Evidence is presented that amounts of fluoride (F) not exceeding one part per million are of no public health significance."

References: 1. *Public Health Repts. Reprint (U.S.)* 1721, Dec. 6, 1935.
2. *Public Health Repts. (U.S.),* **50**: 424 (1936) and APHA, June, 1936.
3. *Public Health Reprint (U.S.)* 1857, Sept. 10, 1937.
4. *Public Health Reprint (U.S.)* 2032, Feb. 10, 1939.
5. *Public Health Repts. (U.S.)* 826, May 26, 1939.
6. *Public Health Repts. (U.S.),* 212, Mar. 3, 1939.
7. *Public Health Repts. (U.S.),* 775, Apr. 11, 1941 (*Reprint* 2260).

and Illinois, Indiana, Ohio, and Virginia. From other sources Dr. Dean learned that similar supplies occurred in the British West Indies, China, Holland, Italy, Mexico, North Africa, South America, Spain, and India. At present the areas in the United States known to contain supplies having more than 0.7 ppm fluoride are shown on the map. There are over 1,200 known communities comprising 4.1 million people in the United States using water containing too much fluoride (that is, in excess of the optimum discussed in Chap. 6). In order to avoid the disfiguring fluorosis which is bound to occur, these places could either change their supplies or obtain a water plant designed to remove the excessive fluorides. Such plants have been built and are described in Chap. 16.

By means of observing many thousands of children who were raised in communities in many parts of the country and who used waters of varying fluoride levels, Dr. Dean established a mottled-enamel index—a numerical method for measuring the severity of fluorosis. With this tool he was able to establish the fluoride level below which the use of such water contributed no significant fluorosis. This level in the latitude of Chicago was about 1.0 ppm.

Many investigators, including Dr. McKay, had observed during the 1920s that there was possibly less decay among children whose teeth were fluorosed. Dr. Dean's next logical step therefore was to verify this observation. Using data already collected and adding many additional children to his lists, he established the second basis for the fluoridation edifice. An examination of many thousands of children in 44 cities in the United States permitted an arrangement of his data somewhat according to Table 1.1. A graph prepared from data (Fig. 1.2) collected during the examinations of 7,257 of these children living in 21 cities clearly and readily shows the remarkable relationship between water-borne fluorides and fluorosis and caries incidence. (A more dramatic graph of a similar study is shown in Fig. 1.3.) Three truths were obvious from Dr. Dean's figures:

1. When the fluoride level exceeds about 1.5 ppm, any further increase does not significantly decrease the decayed, missing, filled (DMF index) tooth incidence (see page 15, Chap. 2), but does increase the occurrence and severity of mottling (Table 1.1).

2. At a fluoride level of about 1.0 ppm the optimum occurs—maximum reduction in caries with no aesthetically significant mottling. He

found that DMF rates were reduced by 60 per cent among the 12- to 14-year-old children (Table 1.2).

3. At fluoride levels below 1.0 ppm some benefits occur, but caries reduction is not so great and gradually decreases as the fluoride levels decrease until, as zero fluoride is approached, no observable improvement occurs.

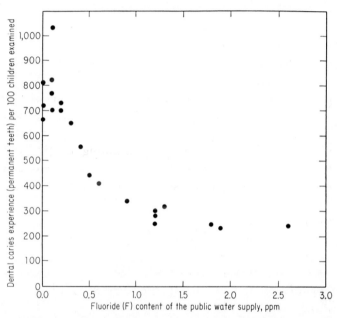

Fig. 1.2. Relation between the amount of dental caries (permanent teeth) observed in 7,257 selected 12- to 14-year-old white school children of 21 cities of 4 states and the fluoride (F) content of public water supply.

With this knowledge it was inevitable that fluorides had to be added to a water supply and the effects measured. It was necessary to determine experimentally whether or not the conditions observed in using naturally fluoridated water could be duplicated when fluorides were deliberately added. This was suggested by Dr. Dean in 1938; by Dr. Gerald J. Cox of the Mellon Institute, Pittsburgh, in 1939; and by Dr. David B. Ast of the New York State Department of Health in 1942. Dr. W. L. Hutton was the first, in 1942, to suggest adding fluorides to a specific water supply—that of Brantford, On-

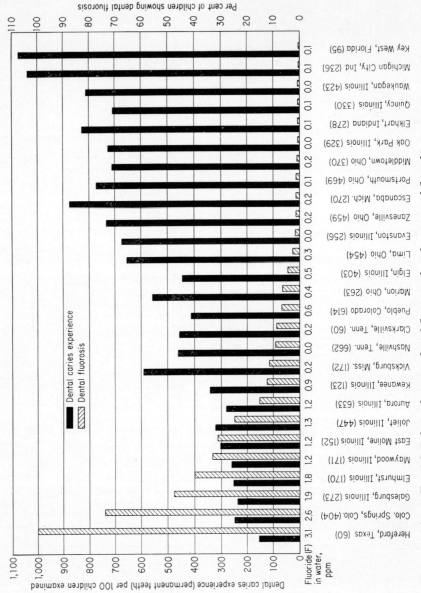

Fig. 1.3. Relation between dental fluorosis, dental caries, and fluorides in water.

TABLE 1.2. DENTAL CARIES EXPERIENCE

City	Number carious permanent teeth per 100 children	Number children examined	Fluorine concentration	Reference
Colorado Springs, Colo	162	203	2.5	1
Galesburg, Ill	194	243	1.9	2
Pueblo, Colo	194	411	0.6	1
Monmouth, Ill	208	99	1.7	2
Elmhurst, Ill	252	170	1.8	3
Maywood, Ill	258	171	1.2	3
Green Bay, Wis	275	687	2.3	1
Aurora, Ill	281	633	1.2	3
Fort Collins, Colo	296	207	0.0	1
Joliet, Ill	323	447	1.3	3
Denver, Colo	342	637	0.5	1
Macomb, Ill	368	63	0.2	2
Elgin, Ill	444	402	0.5	3
Quincy, Ill	628	291	0.2	2
Two Rivers, Wis	646	382	0.3	1
Evanston, Ill	673	256	0.0	3
Manitowoc, Wis	682	661	0.35	1
Sheboygan, Wis	710	244	0.5	1
Oak Park, Ill	722	329	0.0	3
La Crosse, Wis	731	47	0.12	1
Baraboo, Wis	733	119	0.2	1
Waukegan, Ill	810	438	0.0	3
West Allis, Wis	831	160	0.3	1
Milwaukee, Wis	917	2645	0.3	1

References: 1. *Public Health Reprint (U.S.)* 1973, Aug. 19, 1938.
2. *Public Health Reprint (U.S.)* 2073, May 26, 1939.
3. *Public Health Reprint (U.S.)* 2260, Apr. 11, 1941.

tario. The interruption caused by World War II delayed the onset of such demonstrations until 1945, when three cities—Grand Rapids, Michigan, Newburgh, New York, and Brantford, Ontario—began fluoridating.

CHAPTER 2   *Benefits of Fluoridation*

By 1939 it was well established that fluorides in drinking water produced a change in the permanent-tooth enamel so that it became more resistant to dental decay. Although many studies contributed to this result, the basis for the discovery is combined in three separate inquiries:

1. The observations of Dr. G. L. Cox et al.[1] that rat teeth formed during fluoride ingestion have increased resistance to caries
2. The analysis of Dr. W. D. Armstrong et al.[2] showing that carious teeth contained less fluoride than those that had not decayed
3. The report of Dr. Dean et al.[3] that the incidence of caries was smaller among children who had been using naturally fluoridated water than among those using a low-fluoride water

With these studies as a basis, several demonstrations were proposed to fluoridate a public water supply and to observe periodically the effects on children's teeth. In order that the results would be acceptable and conclusive, careful controls had to be incorporated and plans were made during this period to establish such demonstrations.

In Dr. Dean's celebrated "study of 21 cities," described in Chap. 1, it was necessary to devise a means of comparing accurately the dental effects of various fluoride exposures. He did this by establishing certain criteria which were applied equally in each community where he measured these effects.

[1] Fluorine and Its Relation to Dental Caries, *J. Dental Research*, **18:** 481 (1939).
[2] Fluorine Content of Enamel of Sound and Carious Teeth, *J. Dental Research*, **16:** 309 (1937).
[3] Domestic Water and Dental Caries, *Public Health Repts.* (*U.S.*), **54:** 862 (1939).

Benefits of Fluoridation

1. All children examined were born in the community, had lived there continuously (absences up to thirty days per year permitted), and had used the local public water supply during this time.

2. Examinations were made to reveal for each child the number of filled teeth, the number of decayed unfilled teeth, the missing (extracted) teeth, and those which should have been extracted because of extensive tooth decay or other causes.

These two principles of procedure were used in measuring the efficiency of controlled fluoridation at all study projects—not only the one at Grand Rapids, Michigan, but almost all others in the world. Some investigators reported the condition of teeth of children for this purpose as the sum of the decayed, missing, and filled teeth; others reported their findings as a rate—the number of such defective teeth per child, per 100 children, or per 100 teeth examined, or the number of tooth surfaces affected. In any case, and whatever the method of reporting results, this measurement revealed rather accurately the dental-caries experience of the children examined. Eventually most studies reported the improvement in dental health resulting from fluoridation as a reduction in "DMF rates." This term stood for the number of decayed, missing, or filled permanent teeth per child (or per 100 children or per 100 erupted permanent teeth). This figure could be related to any age group comprising either a single year (as, for instance, all 12-year-old children in a community); or it could combine several age groups—those between 6 and 12 years of age, for example.

It was obviously one of the first requirements in such studies to decide on the method by which results could be revealed and compared. The methods used by Dr. Dean in his "21 cities" study (Fig. 2.1) were almost directly applicable in these places, where the fluoride concentrations in water were controlled. Consequently, the techniques of examination, the statistical analysis of the data, and the presentation of results were similar in all studies.

Other methods of measurement of the effects of using fluoridated water were used in addition to the generally accepted DMF rates. These included the so-called "def rates" and "caries-free rates." The first is the number of decayed, extraction indicated, and filled baby teeth (also called the "deciduous" or "milk" teeth). "Caries-free rates" offered a measurement of the increase in the number of children found free of all caries after fluoridation had started.

These methods of measurement were based on a more or less standard method of examinations in all of the early fluoridation studies. Even though the dental examiners at Grand Rapids, Michigan, Newburgh, New York, and Brantford, Ontario, were different individuals, their results, as will be shown, were remarkably similar.

All these studies included a similar examination of children in a control city which was chosen for similarities in every respect except

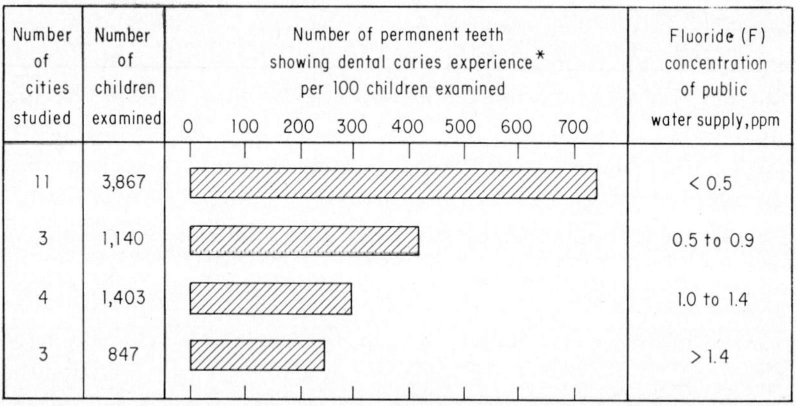

Number of cities studied	Number of children examined	Number of permanent teeth showing dental caries experience* per 100 children examined	Fluoride (F) concentration of public water supply, ppm
11	3,867	▨▨▨▨▨▨ (~650)	< 0.5
3	1,140	▨▨▨▨ (~400)	0.5 to 0.9
4	1,403	▨▨▨ (~300)	1.0 to 1.4
3	847	▨▨ (~200)	> 1.4

* Dental caries experience is computed by totaling the number of filled teeth (past dental caries), the number of teeth with untreated dental caries, the number of teeth indicated for extraction, and the number of teeth missing (presumably because of dental caries)

FIG. 2.1. Dental-caries experience in children of 21 cities.

for the fluoride content of the public water supply. These places, with their original fluoride content (in ppm), were:
1. Grand Rapids, Michigan (0.15 raised to 1.0)
   Muskegon, Michigan (0.15)
   Aurora, Illinois (1.2)
2. Newburgh, New York (0.1 raised to 1.1)
   Kingston, New York (0.1)
3. Brantford, Ontario (0.1 raised to 1.0)
   Sarnia, Ontario (0.1)
   Stratford, Ontario (1.2)

Other studies in the United States which compared results with the base lines established within the studied cities (no control city used) included Marshall, Texas, Evanston, Illinois, Sheboygan, Wisconsin, Ottawa, Kansas, Lewiston, Idaho, and Southbury School, Southbury, Connecticut.

Benefits of Fluoridation

The methods of conducting these studies were very similar; i.e., each year a significant (statistically) number of children in each of the designated age groups were examined in a manner as nearly the same as possible. The number of children examined at Grand Rapids each year is shown in Table 2.1. After 1951 Muskegon could no longer

TABLE 2.1. DISTRIBUTION OF CONTINUOUS-RESIDENT CHILDREN EXAMINED IN GRAND RAPIDS AND MUSKEGON, MICHIGAN, AND AURORA, ILLINOIS, ACCORDING TO AGE AND YEAR OF EXAMINATION†

Age	Aurora, Ill., 1945	Basic examination, 1944–45	1945	1946	1947	1948	1949	1950	1951	1952	1953	1954	1955
						Grand Rapids, Michigan							
4	30	323	540	300	168	137	75	117	168	116	101	77	
5	407	1,633	1,714	831	886	842	777	720	853	1,087	715	529	
6	473	1,789	1,186	628	663	736	697	748	750	826	1,010	561	
7	516	1,806	149	82	69	55	54	438	423	422	410	751	
8	469	1,647	15	216	135	138	155	501	470	444	390	567	
9	368	1,639	.....	525	465	484	519	520	582	720	623	477	
10	397	1,626	.....	109	108	111	125	131	141	512	499	515	
11	383	1,556	.....	17	18	22	140	130	151	246	291	499	27
12	401	1,685	174	85	38	60	130	200	176	211	316	260	168
13	401	1,668	953	547	625	600	574	530	497	497	557	224	254
14	433	1,690	273	173	196	152	153	130	128	119	111	250	254
15	467	1,511	80	53	80	64	64	58	53	80	99	240	307
16	371	1,107	4	3	233	245	209	177	198	191	197	198	212
Total....	5,116	19,680	5,088	3,569	3,684	3,646	3,672	4,400	4,590	5,471	5,319	5,148	1,222
						Muskegon, Michigan‡							
4	.....	20	.....	43	18	26	51	41	63	52	43	40	
5	.....	402	.....	321	348	422	340	359	351	487	370	381	55
6	.....	462	.....	339	312	305	393	310	294	353	397	386	72
7	.....	408	.....	36	42	36	30	274	223	246	209	292	78
8	.....	376	.....	18	13	10	12	190	275	205	212	244	75
9	.....	357	.....	213	215	199	197	227	277	348	258	275	65
10	.....	359	.....	62	57	52	52	51	62	287	311	226	63
11	.....	293	.....	12	10	14	146	141	139	133	175	208	54
12	.....	328	.....	21	19	11	28	43	48	46	163	183	93
13	.....	377	.....	197	207	208	214	173	225	178	228	243	58
14	.....	369	.....	77	50	79	66	63	59	54	51	121	
15	.....	292	.....	18	44	41	34	35	21	30	35	139	
16	.....	248	.....	1	199	205	132	146	155	132	161	185	
Total....	.....	4,291	.....	1,358	1,534	1,608	1,695	2,053	2,192	2,551	2,613	2,923	613

† Francis A. Arnold, Jr., Grand Rapids Fluoridation Study—Results Pertaining to the Eleventh Year of Fluoridation, *Am. J. Public Health*, **47**(5): 539 (1957).

‡ The basic examinations in Muskegon were not done until late spring of 1945; therefore, no examinations were made in the fall of 1945.

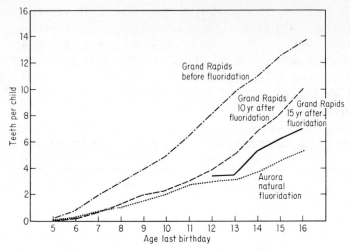

Fig. 2.2. Permanent teeth DMF reduction at Grand Rapids after ten and fifteen years of fluoridation. (*USPHS.*)

be continued as a control city because fluoridation of its public water supply began then.

As a result of these techniques it was possible to follow the progress of fluoridation results very closely. A typical report of this progress is shown in Fig. 2.2. The base lines showing the initial DMF rates at Grand Rapids, Muskegon, and Aurora are shown in Fig. 2.3.

The results of other study projects showed similar DMF reduc-

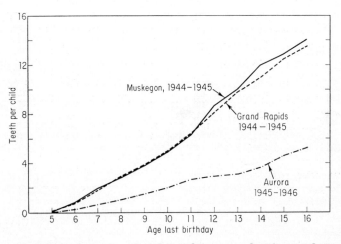

Fig. 2.3. DMF reductions at Grand Rapids, compared with control cities.

Benefits of Fluoridation

tions. Figure 2.4 compares the ten-year results of Grand Rapids,[4] Newburgh,[5] and Brantford.[6] In spite of the differences in the water supplies, climates, dentists who made the examinations, and natural differences in the children examined, the results are obviously very similar.

It can be readily seen that from these curves and charts several conclusions can be drawn:

1. Each successive yearly examination revealed a progressive improvement in the reduction of decayed, missing, and filled teeth.

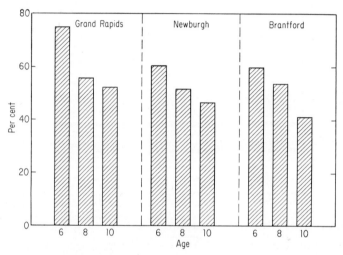

FIG. 2.4. Reduction in tooth decay at pioneer cities.

2. The younger age groups showed the earliest improvement and at any given time the greatest DMF-rate reduction (Fig. 2.5).

3. As the duration of the fluoridation procedure approached the age of the children, the DMF rates approached the rate of Aurora (where the water supply contains naturally 1.2 ppm fluoride), which

[4] Francis A. Arnold, Jr., H. Trendley Dean, Philip Jay, and John W. Knutson, Effect of Fluoridated Public Water Supplies on Dental Caries Prevalence, *Public Health Repts. (U.S.)*, **71**(7): 652–658 (1956).

[5] David B. Ast, David J. Smith, Barnet Wachs, and Katherine T. Cantwell, Newburgh-Kingston Caries-Fluorine Study. XIV. Combined Clinical and Roentgenographic Dental Findings after Ten Years of Fluoride Experience, *J. Am. Dental Assoc.*, **52**: 314–325 (1956).

[6] William L. Hutton, Bradley W. Linscott, and Donald B. Williams, Final Report of Local Studies on Water Fluoridation in Brantford, *Can. J. Public Health*, pp. 89–91, March, 1956.

was considered as the objective and as the ultimate in DMF reduction with this procedure. In other words, the younger the child when fluoridation starts, the better the results.

4. Children whose permanent teeth were already present in the mouth at the time fluoridation started derived a measurable benefit. For instance, the 16-year-old children who were 6 at the start of fluoridation showed a 26 per cent DMF reduction in ten years.[7]

The 16-year-old age group was the oldest examined during the Grand Rapids study. As shown in Fig. 2.6, after twelve years of fluoridation, this group showed a 34 per cent reduction in DMF rates.

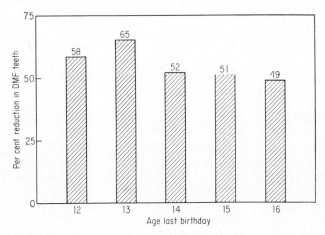

Fig. 2.5. Reduction in tooth decay in specific age groups 15 years after fluoridation at Grand Rapids, Michigan. (*USPHS.*)

This would indicate that young adults derived some benefits from fluoridation even though most of their permanent teeth had been formed or had erupted when fluoridation started. There is every reason to believe that had the opportunity been taken to examine older age groups, a similar (but obviously diminished) improvement in the resistance to dental decay would have occurred.

For instance, there is no reason to believe that DMF reduction in the 17- or 18-year-old group should abruptly fall to zero per cent from the 34 per cent reduction found for the 16-year-old group. Similarly, if even older young adults had been examined, there probably would have been a measurable reduction in DMF teeth which would prob-

[7] Arnold, Dean, Jay, and Knutson, *op. cit.*

Benefits of Fluoridation

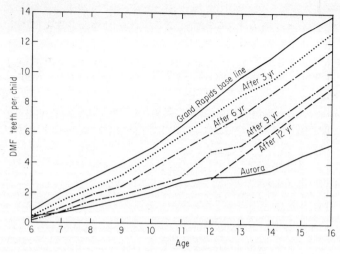

Fig. 2.6. Comparison of DMF experience in Grand Rapids and Aurora.

ably extend into the groups more than 20 years of age. No studies have as yet been made to prove this suspected benefit to older adults.

Deciduous teeth (also called "baby" or "milk" teeth) showed a decided reduction in the decay rates. As shown in Fig. 2.7, the decay rate (def rate) for Grand Rapids after ten years of fluoridation approached that of the experience with baby teeth at Aurora (with 1.2 ppm fluoride naturally) among the youngest children examined and remained unchanged thereafter. The cause of this rather unex-

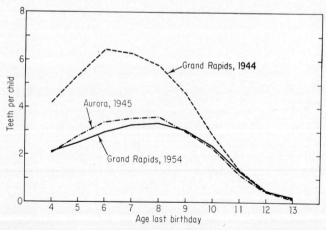

Fig. 2.7. Reduction in tooth decay in deciduous teeth at Grand Rapids and Aurora. (*USPHS.*)

TABLE 2.2. GRAND RAPIDS FLUORIDATION STUDY
CONTINUOUS RESIDENTS
Permanent Teeth

Grand Rapids

Age	Number examined 1955 after 11 years	Number examined 1956 after 12 years	Per cent reduction in DMF teeth 1955 after 11 years	Per cent reduction in DMF teeth 1956 after 12 years	Number DMF teeth per child 1955 after 11 years	Number DMF teeth per child 1956 after 12 years	Per cent with DMF teeth 1955 after 11 years	Per cent with DMF teeth 1956 after 12 years	Missing teeth per child 1955 after 11 years	Missing teeth per child 1956 after 12 years	Filled teeth per child 1955 after 11 years	Filled teeth per child 1956 after 12 years	Carious teeth per child 1955 after 11 years	Carious teeth per child 1956 after 12 years
11	27	30	55.5	44.9	2.852	3.533	66.7	90.0	0.037	0.433	2.074	1.900	0.926	1.200
12	168	160	58.3	63.9	3.363	2.913	81.0	78.8	0.101	0.131	2.554	1.994	0.851	0.900
13	254	265	50.1	54.1	4.858	4.468	88.6	87.9	0.445	0.283	3.098	3.057	1.413	1.600
14	254	330	44.0	45.0	6.130	6.021	93.7	92.1	0.598	0.579	4.031	4.133	1.650	1.473
15	307	331	37.1	39.8	7.857	7.520	95.8	95.8	0.798	0.819	5.241	5.039	2.068	1.864
16	212	272	33.9	34.3	8.929	8.871	97.6	97.8	0.929	0.879	6.316	6.562	1.901	1.787
17	131	129			11.397	9.791	99.2	98.4	1.389	1.209	8.718	7.140	1.634	1.690
18		27				10.074		100.0		2.111		7.037		1.148
Total	1,353	1,544												

Muskegon

Age	Number examined After 4 years	Number examined After 5 years	Per cent reduction in DMF teeth After 4 years	Per cent reduction in DMF teeth After 5 years	Number DMF teeth per child After 4 years	Number DMF teeth per child After 5 years	Per cent with DMF teeth After 4 years	Per cent with DMF teeth After 5 years	Missing teeth per child After 4 years	Missing teeth per child After 5 years	Filled teeth per child After 4 years	Filled teeth per child After 5 years	Carious teeth per child After 4 years	Carious teeth per child After 5 years
5	55	81	100.0	81.5	0.000	0.012	0.0	1.2	0.000	0.000	0.000	0.000	0.000	0.012
6	72	58	84.6	95.8	0.125	0.034	6.9	3.4	0.000	0.000	0.083	0.000	0.042	0.034
7	78	77	57.4	65.4	0.846	0.688	38.5	29.9	0.026	0.000	0.551	0.429	0.269	0.234
8	75	72	46.9	39.2	1.493	1.708	54.7	58.3	0.067	0.028	1.013	1.097	0.413	0.583
9	65	69	26.1	45.6	2.815	2.072	76.9	73.9	0.200	0.130	2.046	1.493	0.646	0.464
10	63	62	27.8	31.9	3.540	3.339	82.5	87.1	0.270	0.306	2.587	2.500	0.746	0.613
11	54	51	40.5	32.3	3.759	4.275	90.7	94.1	0.241	0.388	2.481	3.059	1.130	0.765
12	93	12	44.2	48.0	4.828	4.500	95.7	83.3	0.226	0.083	4.011	4.083	0.688	0.333
13	58		33.2		6.672		96.6		0.638		5.017		1.121	
Total	613	482												

Deciduous Teeth

Grand Rapids

Age	Number examined		Number def teeth per child		Per cent with def teeth		Filled teeth per child	
	1955 after 11 years	1956 after 12 years	1955 after 11 years	1956 after 12 years	1955 after 11 years	1956 after 12 years	1955 after 11 years	1956 after 12 years
11	27	30	0.704	0.533	33.3	26.7	0.407	0.367
12	168	160	0.470	0.589	23.8	22.5	0.214	0.275
13	254	265	0.157	0.166	8.3	10.9	0.055	0.057
14	254	330	0.087	0.055	5.9	3.9	0.031	0.018
15	307	331	0.016	0.030	1.3	2.4	0.003	0.015
16	212	272	0.005	0.007	0.5	0.7	0.005	0.004
17	131	129	0.031	0.008	2.3	0.8	0.031	0.008
18	....	27						
Total	1,353	1,544						

Muskegon

	After 4 years	After 5 years	After 4 years	After 5 years	After 4 years	After 5 years	After 4 years	After 5 years
5	55	81	2.855	2.790	69.1	61.7	1.291	0.988
6	72	58	4.250	3.414	81.9	77.6	1.625	1.879
7	78	77	5.090	4.662	84.6	83.1	2.654	2.247
8	75	72	5.080	5.139	90.7	88.9	2.973	2.972
9	65	69	3.800	3.565	83.1	87.0	2.308	2.014
10	63	62	1.937	2.613	60.3	72.6	0.698	1.419
11	54	51	1.537	0.902	53.7	41.2	0.593	0.333
12	93	12	0.484	0.667	26.9	33.3	0.140	0.250
13	58	....	0.224	....	12.1	....	0.086	....
Total	613	482						

pected benefit cannot be explained on the basis of fluoride absorption by the tooth enamel through the blood stream because deciduous teeth are essentially completely formed at birth. This phenomenon of def-rate improvement might be explained as an absorption mechanism by the deciduous teeth after eruption, the fluoride being incorporated in the enamel structure from the water as it touches the teeth during drinking. In other words, the only source and entry of fluorides left is in the fluoridated drinking water, the other two possible portals having been eliminated; namely, fluoride absorption through the system during the period of tooth-enamel formation and the virtual absence of placental transfer of fluorides (obtaining fluorides from the mother before birth). It has been shown that exposure of pregnant women to fluoridated water maintained at the optimum level has no observable effect on the caries experience of the deciduous teeth of their children.[8]

Except for the effect on the teeth, no other changes of any sort, adverse or beneficial, have ever been found as a result of fluoridated water used in the amounts recommended. This conclusion, of course, had been observed many thousands of times among individuals using water containing naturally many times the fluoride content now considered optimum. In order to confirm this, however, where the fluoride levels were controlled, some of the fluoridation projects included medical and dental observations of varying complexities in addition to those related to the computation of DMF or def rates. These examinations included fluorosis prevalence (mottling of the teeth), X-ray examinations of various joints and parts of the spine, laboratory tests of the blood and urine, and thorough clinical examinations of the children in the group (as at Newburgh, New York).

The observations which were made in the areas containing water supplies with natural fluorides, and which related to fluorosis incidence, were confirmed in the fluoridation-study projects where the fluoride levels were controlled. At Grand Rapids, for instance, it was found that approximately 4 per cent of the children 11 years of age had a very mild degree of mottling, whether or not caused by the fluorides in the water.[9] This degree of mottling is practically not

[8] James P. Carlos, Alan M. Gittelsohn, and William Haddon, Jr., Caries in Deciduous Teeth in Relation to Maternal Ingestion of Fluoride, *Public Health Repts.* (*U.S.*), **77**: 658 (1962).

[9] Francis A. Arnold, Jr., Grand Rapids Fluoridation Study—Results Pertaining to the Eleventh Year of Fluoridation, *Am. J. Public Health*, **47**(5): 539 (1957).

noticeable and is of no aesthetic or public health significance; in fact, it has to be searched for quite diligently by dentists who are experienced in making such examinations and aware of this kind of mottling. In appearance it is a slight, white, opaque flecking of a portion of no more than two or three of the side and rear teeth.

It is this complete absence of any ordinarily noticeable fluorosis that provides the best indication of the constant rate of water consumption among children. If several instances of even mild fluorosis in any of the controlled fluoridation studies had occurred, it would have been presumed that such children had consumed an excessive amount of water during the critical period of their tooth-enamel formation. Actually, not a single such instance has so far been observed. In other words, even though there is undoubtedly a difference in the amounts of water children drink, the variation is not large enough to produce a noticeable fluorosis from the fluorides contained in their drinking water. This is believed to be one of the most significant arguments for using drinking water as the medium or carrier of the fluoride required for optimum dental health. No other suggested media are believed to have the same pattern of constant, relatively unvarying consumption as has been demonstrated for water.

In regard to *all* other physical examinations and observations in *all* of the studies so far reported, a typical conclusion is: "No differences of medical significance could be found between the two groups of children."[10]

[10] Edward R. Schlesinger, David E. Overton, and Helen C. Chase, Study of Children Drinking Fluoridated and Non-fluoridated Water, *J. Am. Med. Assoc.*, 160: 21–24 (1956).

CHAPTER 3 *Adoption Procedures*

More than 17 years have elapsed since the start of the first fluoridation projects at Grand Rapids, Michigan, Newburgh, New York, and Brantford, Ontario. The data included in Chap. 2 confirm the effectiveness of these programs and show that the procedure is safe, convenient, and most economical. As a result, many other communities have adopted the measure, so that at present more than 2,000 communities are adding fluorides to their supplies in controlled amounts. As will be shown in Chap. 14, most of the communities that have adopted fluoridation are those with the largest populations; the large number of places without fluoridation and with populations of less than 2,500 indicates that for various reasons many people have not considered it worthwhile to adopt the measure.

In view of the attractive economies and other advantages to communities where fluoridation has been adopted, it may eventually be the task of all public health and city government officials to explain the advantages of this measure to their associates and constituents. The fact that many people have not been informed of these advantages was brought out in a poll conducted by Elmo Roper, who found that almost 30 per cent of the people asked had never heard of fluoridation, even though another 52 per cent thought it was a good thing. Despite this latter figure, more fluoridation referendums have been defeated than have been adopted.

The measure should be considered seriously in every community where the water supply is deficient in fluorides because in the United States at the present time:

Over 90 per cent of the 7-year-old children have one or more decayed teeth.

Among all the 16-year-old children an average of seven teeth have become carious.

The average adult has only one-half his teeth by the age of 40. Over 21 million persons have lost all of their teeth.

There are three factors which aggravate this problem and make it difficult to solve:

1. There is widespread indifference to oral health problems.
2. There are only about one-half as many dentists as are considered adequate to maintain proper dental care.
3. The cost of adequate dental care is beyond the means of many families.

It has been shown that one of the best ways of alleviating these dental prospects is to make fluoridated water available to everyone.

In order that the measure may be adopted in your community, it is prudent to study the causes of failure in other areas and to be prepared to cope with them. Moreover, it is even more advisable to know what kind of organization and procedures were successfully applied where fluoridation was adopted. A somewhat typical procedure, which has been used with many modifications in a wide variety of communities, is outlined in Fig. 3.1.

In this procedure, the idea for fluoridating the water supply of the community is probably first brought to the attention of the dentists or physicians through their technical journals, conventions, or other means. They approach their local dental or medical societies, which endorse the measure for their community with little, if any, delay. From the scientific point of view, fluoridation is the least controversial public health measure we have ever had. Those who have studied the results and the background leading to these results, and are trained to interpret them, have learned that fluoridation substantially decreases dental decay with safety and efficiency. As a result, the measure is endorsed by every related professional and scientific organization, every state health department, and a host of lay organizations having an interest in community betterment. A partial list of such groups is included at the end of this chapter.

The dental or medical society then appoints a health advisory group representing the professional health groups, the dental and medical societies, and the voluntary health agencies. Their function is to provide guidance and policies and to furnish speakers for lectures at public meetings. From this group a coordinator or chairman is appointed.

The guidance for the entire campaign rests with this individual, who should be chosen with considerable care. An evaluation of his qualifications should include such factors as (1) the standing he has attained in the community, (2) his occupation or profession, (3) the

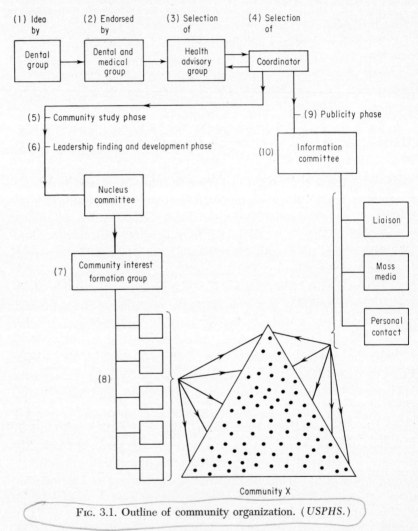

Fig. 3.1. Outline of community organization. (*USPHS.*)

length of time he has lived in the community, (4) his record of other successful community leadership functions, and (5) his interest in the improvement of his community.

Under such leadership the health advisory group makes a study of the community from the point of view of (1) the feasibility and cost of fluoridating their public water supply (such information would come primarily from the city engineer or water-plant superintendent); (2) the political factors bearing on the measure; e.g., local opinion of water-plant operation, local satisfaction with other public utilities; (3) characteristics of the community; e.g., proportion of people over 45 years of age, number of families with young children, attitudes of people to new, progressive measures, unanimity of aim of city council members; (4) history of similar measures which have successfully passed; e.g., schools, treatment works, parks, streets and roads, and the like.

This survey provides some idea of the problems to be overcome. The advisory group is now ready to appoint leaders to work in various parts of the city or with specialized groups. They will be informed as to possible objections. Every effort will be made to get information as to the advantages of fluoridation to every person in the community.

It will be relatively easy to present the advantages to be gained in your community from fluoridation if you base them on the experiences of other cities. The most difficult problem will be the task of answering the questions and allegations raised by the opponents of fluoridation.

Opposition to fluoridation is in most instances not limited to individuals in the community. There are several nationwide organizations composed of small but vocal groups which stir up and lead the opposition. The methods are basically similar to those used in the past against other health measures, such as vaccination, pasteurization, and chlorination. The tactics used by the opponents spread fear and doubt about the value of fluoridation. It is therefore essential that intense efforts be made to inform the community of the benefits of fluoridation which are available at little cost and without harmful results to anyone.

With a knowledge of their opponents' tactics, the neighborhood leaders will be able to answer any opposition effectively. If properly conducted, the campaign will result in a demand for fluoridation which will be satisfied by a resolution, the passage of an ordinance by the governing body of the community, or the favorable adoption of a referendum on fluoridation.

With the legal authorizations obtained, the plans for fluoridating the water supply are obtained from either the city's own engineering forces or from an outside consulting engineer. These plans, along with cost estimates and the resolutions of the local medical and dental societies, are sent to the state health department, which will authorize the city to proceed after reviewing and approving these plans. Money is then appropriated and the equipment is purchased and installed. Another community is then added to the ever-growing list of those that have elected to preserve, as far as is presently possible, their children's teeth.

In this country fluoridation has now been approved by every major scientific and professional organization which is competent in this field. Approval has also been given by many related organizations, all state health departments, and by responsible health, medical, and dental officials throughout the world. Among the organizations which have formally approved or endorsed the measure are:

American Academy of Pediatrics
American Association for the Advancement of Science
American Association of Public Health Dentists
American College of Dentists
American Dental Association
American Dental Hygienists Association
American Federation of Labor and Congress of Industrial Organizations
American Hospital Association
American Legion
American Medical Association
American Nurses Association
American Pharmaceutical Association
American Public Health Association
American Public Welfare Association
American School Health Association
American Society of Clinical Pathologists
American Society of Dentistry for Children
American Veterinary Medical Association
American Water Works Association
Association of Casualty and Surety Companies
Association of State and Territorial Dental Directors
Child Study Association of America
College of American Pathologists
Commission on Chronic Illness

Adoption Procedures

- Conference of State Sanitary Engineers
- Federal Civil Defense Administration
- Industrial Medicine Association
- Joint Committee on Health Problems of the American Medical Association and the National Education Association
- Lions International
- National Congress of Parents and Teachers
- National Research Council
- State and Territorial Health Officers Association
- U.S. Department of Defense
- U.S. Department of Health, Education and Welfare: U.S. Public Health Service
- U.S. Junior Chamber of Commerce
- World Health Organization

In addition, the U.S. Department of Defense has issued a number of policy statements covering the subject of fluorides in the water supplies serving various military bases and posts. A typical example is this statement from the Department of the Army (ACAC-C (M)671.1-MEDCA, 7/1/54): "Use of this procedure is recommended at posts where the natural fluorine content of the post water supply is very low and where there is a substantial child population."

CHAPTER 4   *Fluoridation Litigation*

For as many reasons as there are facets to human nature, opposition to fluoridation has developed and continues to retard its adoption in many communities. Some of these reasons are quite valid and are worthy of serious consideration and rebuttal. The questions relating to legality have practically invariably been answered in favor of fluoridation. The litigation so far has followed the same pattern in relation to one or more of the following five contentions:

1. That no reasonable relationship exists between fluoridation and the maintenance of public health
2. That the prevention of dental decay is not a proper objective of joint community effort
3. That fluoridation cannot be legally authorized by a community
4. That fluoridation violates constitutional rights (such as religious freedom and other fundamental liberties)
5. That fluoridation represents the unlicensed practice of medicine, dentistry, and pharmacy

The first three of these objections relate to "police power" or the legal right of a community to authorize fluoridation. This fundamental question was one of the first legal considerations brought before a court.

It has been averred that the legislative action authorizing fluoridation is not a valid exercise of police power of a community and is therefore unlawful and unconstitutional. An understanding of police power can perhaps be best presented by quoting Mr. Justice Harlan in an opinion on the validity of a statute on vaccination:

> The authority of the State to enact this statute is to be referred to what is commonly called the police power—a power which the State did not surrender when becoming a member of the Union under the

Constitution. Although this court has refrained from any attempt to define the limits of that power, yet it has distinctly recognized the authority of a State to enact quarantine laws and health laws of every description. . . . According to settled principles, the police power of a State must be held to embrace, at least, such reasonable regulations established directly by legislative enactment as will protect the public health and the public safety. . . . It is equally true that the State may invest local bodies called into existence for the purposes of local administration with authority in some appropriate way to safeguard the public health and the public safety. The mode or manner in which those results are to be accomplished is within the discretion of the State, subject, of course, so far as Federal power is concerned, only to the condition that no rule prescribed by a State, nor any regulation adopted by a local government agency acting under the sanction of State legislation, shall contravene the Constitution of the United States or infringe any right granted or secured by that instrument.

From this and similar opinions it is clear that a state may adopt such legislation as it believes necessary to promote the public health and safety. It may also delegate in part certain legislative powers in these areas to local municipal bodies or health departments. This delegated power has certain limitations when exercised by a community, and hundreds of court cases have established definite boundaries to these powers. It is generally recognized, however, that a public health enactment is a valid exercise of police power when (1) its purpose is reasonably related to the maintenance and protection of the public health; (2) if otherwise justified, the means, rules, or regulations adopted are not an arbitrary, unreasonable, or oppressive invasion of personal rights secured by the Federal or state constitutions; or (3) if otherwise justified, the exercise of this power does not conflict with overriding Federal or state laws.

All three of these limitations have been cited as being violated when communities have authorized fluoridation of their water supplies. The courts have ruled in every case that such authorization is a proper utilization of the communities' police power. In cases relating to the first of these limitations, i.e., that the purpose of fluoridation is related to the maintenance of public health, the courts have ruled that fluoridation is in fact reasonably related to the protection of the public health because its purpose is primarily to reduce dental disease among a substantial portion of the total population although the disease thereby reduced is not communicable or contagious. The Supreme Court of Washington, for instance, ruled:

Protection of public health includes protection from the introduction or spread of both contagious and non-contagious diseases. There is a direct and significant relationship between dental health and general bodily health of individuals. We find nothing in this jurisdiction which limits the police power, exercised in the realm of public health, solely to the control of contagious diseases, as distinguished from non-contagious diseases. Further, under the police power, a health regulation may be effective police measure, without the existence of some immediate public necessity.

Further, the Supreme Court of Louisiana stated:

Their (the children of a community) health and physical well-being is of great concern to all the people, and any legislation to retard or reduce disease in their midst cannot and should not be opposed on the ground that it has no reasonable relation to the general health and welfare. Children of today are adult citizens of tomorrow upon whose shoulders will fall the responsibilities and duties of maintaining our government and society. Any legislation therefore which will better equip them by retarding or reducing the prevalence of disease is of great importance and beneficial to all citizens.

The most important issue in almost all the cases against fluoridation has been the possible interference with fundamental personal rights by an otherwise valid exercise of the police power. Such personal rights have included:

1. The right of each individual to treat his health as he pleases
2. The right of a parent to safeguard the health of his children as he pleases
3. The right of each individual to be free of medical experimentation
4. The right to be free of arbitrary, unreasonable, or oppressive invasions of personal prerogatives
5. The right of each individual to be free of interference with his religious beliefs

The fundamental rights of individuals are guaranteed under the First and Fourteenth Amendments of the Constitution of the United States. The relevant portions of these amendments are:

ARTICLE I: Congress shall make no law respecting an establishment of religion, or prohibiting the free exercise thereof; . . .

ARTICLE XIV. SEC. 1: . . . No State shall make or enforce any law which shall abridge the privileges or immunities of citizens of the United States; nor shall any State deprive any person of life, liberty, or property, without due process of law; nor deny to any person within its jurisdiction the equal protection of the laws.

This section of the Fourteenth Amendment protects the liberties of an individual against an action taken by a state or community without due process of law. Accordingly, any of the personal liberties secured under the Fourteenth Amendment is subjected to restraint and qualifications imposed by a reasonable and legitimate exercise of police power (due process of law). In all the cases in which fluoridation is alleged to deprive an individual of his right to protect his and his children's health without interference, the courts have ruled that the police power delegated to a community to promote the public health is predominant. Any legal action or ordinance to permit fluoridation does not compel an individual to do anything nor does it subject him to any penalty. Liberty implies absence of arbitrary restraint. It does not imply immunity from reasonable regulations adopted to promote the health and well-being of the community.

All the other personal liberties mentioned in cases involving fluoridation have been decided by similar reasoning with the exception of those involving the question of freedom from interference with religious beliefs. Such pleas have been considered more seriously and extensively than any other. Those people who hold religious scruples against the use of medicines believe that fluoridation deprives them of their constitutional right of freedom of religion by forcing on them a medicinal treatment.

The courts have many times, however, interfered with the free practice of religion, especially where such action would affect adversely the public health or, in the case of polygamous marriages, offend the principles upon which a community is founded. Cases involving vaccinations, blood transfusions, chlorination of water, and submission to X rays and blood tests have been decided in favor of the community even though such actions were at variance with complete religious liberty. In ruling on this question, brought by the city of Oroville, California, the court stated:

> We recognize that it is a fundamental constitutional principle that a person is entitled to adhere to any religious belief which he may choose. However, there is another principle which is equally true and fundamental—that no person may, by exercising his religious belief, infringe the sovereign power of the State to provide for the health, safety, or general welfare of its citizens. When these two principles

collide, the power of the State must prevail. The Supreme Court of the United States has laid down the basic rule that the right to *think* and *believe* is unlimited but that the right to *act* in pursuance of such thought or belief is, necessarily, *limited*. The inadmissible position of these protestants is that the customers of this defendant utility must be denied the benefit derived from the fluoridation of the water supply of said utility because certain customers assert that fluoridation infringes upon their constitutional right to religious freedom.

In addition to these commonly contested actions, a large number of claims and assertions have been considered and decided on by various courts. Among these are the following.

It has been claimed that fluoridation is arbitrary because it benefits only a small portion of the population; viz., children. The court declared that eventually all the people will have benefited as such children reach maturity. A police measure designed to protect the public health is not arbitrary merely because not everyone is affected.

The courts have frequently discussed the claim that fluoridation is mass medication and that a community should not force a medication on everyone except to control a contagious or communicable disease. The courts have answered: first, that no one is compelled to drink the water; and second, that the addition of fluorides to water is not strictly medication in the usual sense. The city is in fact adding a mineral to a water which has been found to be deficient in that mineral. In this connection the distinction is usually drawn between chlorination and fluoridation; i.e., the treatment of water and the treatment of people. The courts have held that the distinction is immaterial whether, in attempting to improve the public health, bacteria are killed or teeth are hardened against decay.

Related to the question of medication is the contention that when a community authorizes fluoridation, it is in reality engaged in the illegal practice of medicine, dentistry, or pharmacy. The courts have stated that a community adopting fluoridation is no more engaging in such a practice "than a mother would be who furnishes her children a well-balanced diet. . . . "

A considerably more serious objection has been raised when the "right" of a community to authorize fluoridation has been questioned. In some cases, the ordinance providing for fluoridation may be in excess of the power granted the community in its charter. Usually such charters give the city the right to enact ordinances to secure the public health, safety, and welfare. If there are no such rights conferred in a charter, the same authority may be contained in laws establishing state boards of health, public utility commissions, or various boards of inspections or licensure. So far, no community has been found to be without such legal right to adopt fluoridation if they so wish.

It has been contended that fluoridation of a community water supply is the equivalent of food adulteration and therefore a violation of the Federal and state pure-food laws. Under the Food, Drug and Cosmetic Act, food is defined as: "Articles used for . . . drink for man or other animals." State laws usually provide that drinking water is to be inspected and approved by a state authority, usually the state health department. If they so approve a fluoridated water supply as being fit for human consumption, then fluoridated water would not ordinarily come under such pure-food laws. The Federal Food and Drug Administration has stated that they will "regard water supplies containing fluorine, within the limitation recommended by the U.S. Public Health Service, as not actionable" under their laws. So far, no court has found that the provision of fluoridation was in conflict with the state pure-food laws.

Many who have questioned the right of a community to authorize fluoridation have sought to prove that the measure is not effective or that its safety has not been proved sufficiently for universal adoption, or adoption at least in that particular community. Such questions are not within the purview of a court; such findings have had to be made by the legislative or administrative body prior to the introduction of the case into a court. The courts have had to determine, however, whether the effectiveness and safety of fluoridation as determined by the communities' governing bodies are actually supported by competent evidence. Such cases have accumulated a large body of evidence and testimony on the effects of fluorides on the various organs and tissues of man and animals and on the teeth of persons exposed to various concentrations of fluoride in water. In every case, sufficient competent evidence has been found to support the administrative and legislative bodies' determination that fluoridation is in fact safe and effective and will not in any way imperil the health and well-being of any segment of the communities' population.

There is evidently no difference in the mode of compliance with a resolution, ordinance, or order to begin fluoridation issued by a municipal legislative body for either a municipally or a privately owned water company. A water system operated by a community would be acting as an agent in complying with the order issued under the police power. A private water company similarly acts as an agent of the same municipal legislative body. At the present time over two hundred communities that have adopted fluoridation are served by private water companies.

One other legislative consideration involves the liability of a municipality in the event someone is harmed by any aspect of the fluoridation procedure. This has never happened, but it is generally believed that in such a case the municipality would be judged from the standpoint of what negligence, if any, could be proved.

Up to the present, the courts in ten states have held that none of the various pleas mentioned in this chapter were valid. To strengthen their decisions, the United States Supreme Court has refused to review four of the decisions because no Federal constitutional question was involved. In only one instance was an adverse opinion obtained, and this was promptly reversed by the Supreme Court of Louisiana.

More detailed discussions of this subject and of some of the cases involved can be found in the following publications:

1. William P. Stallsmith, Jr., Legal Aspects of the Fluoridation of Public Drinking Water, *George Washington Law Rev.*, **23**(3): 343 (1955).

2. John H. Murdock, Jr., Legal Aspects of Water Fluoridation, *Water and Sewage Works*, Sept. and Nov., 1956.

3. American Dental Association, Council on Legislation.: A Review of Court Decisions on Fluoridation of Public Water Supplies, *J. Am. Dental Assoc.*, **3**: 379–380 (1957).

4. B. J. Conway, Legal Aspects of Municipal Fluoridation, *J. Am. Water Works Assoc.*, **50**(10): 1330 (1958).

5. Anon., Master's Report in the Chicago Fluoridation Suit, *Pure Water*, **13**(11 and 12): 144 and 167 (1961).

CHAPTER 5 *Technical Objections (Engineering, Chemical, Industrial, Economic)*

The large number of arguments advanced by the opponents of fluoridation which are related to the engineering, chemical, and industrial aspects can be grouped under these headings:
1. Chemical compounds and theories
2. Problems arising in the water plant because of fluoridation
3. Commercial and industrial use of fluoridated water
4. Other methods for administering fluorides

CHEMICAL COMPOUNDS AND THEORIES

While many of the opponents of fluoridation have admitted the validity of the findings showing the improved dental health among people living in areas where fluorides occur naturally in the water, they doubt whether the same results can be achieved in communities with controlled fluoridation. Their doubt is based on their belief that there is a fundamental difference between "natural" and "artificial" fluorides. Many outstanding chemists have denied that such a difference can exist. A group of such chemists has stated:

> The element *fluorine* is made up of atoms that have a definite structure. When fluorine combines with other elements, or group of elements, each atom gains one electron and the new substance is called a fluoride. In water solution these fluoride particles tend to dissociate as separate charged particles called ions. The fluoride ions have different properties than the element fluor*ine*, just as chlor*ine* gas has different properties than chlorides such as table salt. The terms "artificial" or "chemical" are sometimes used in a misleading sense and cause

confusion with reference to the use of refined or prepared fluorides for regulating the composition of water supplies. Since the fluor*ide ion* dissolved in water, either above or below ground, is the same as the fluoride ion in water when fluorides are added under good chemical and engineering control to prevent dental caries, there will be no difference in their effect in the human body.[1]

The intense manner with which fluorine reacts with other elements is often cited as a reason for avoiding fluoride compounds for consumption. In this case, elemental fluorine [which is never used for fluoridation (see Chap. 7)] is confused with the compounds of fluorine, which are quite docile. A similar situation occurs in the union of the very reactive elements comprising ordinary table salt—sodium and chlorine. That the introduction of fluoride compounds into a water supply is not harmful, however, is borne out by the large number of reports that the benefits of both artificially and naturally fluoridated waters are similar.

It is claimed that the compounds used for fluoridation are not pure, not "organic," and hence not natural or beneficial. All the compounds now in use are described in Chap. 7. Many are over 99 per cent pure; the others contain varying amounts of predominately harmless impurities (fluosilicic acid, for instance, contains more than 70 per cent water). Some people claim that fluorides are a waste product of the aluminum industry. The aluminum industry, in fact, uses large quantities of fluoride compounds and avoids disposing of them because they must be provided as a raw material for making aluminum. Fluorides are certainly not a by-product of the aluminum industry.

Fluoridation has been cited as a wasteful procedure inasmuch as only children benefit. It is true that fluorides in water are of benefit only to children within a certain age group (0 to about 16 years of age). The benefits derived, however, last throughout their entire lives and children generally become adults. Then too, the cost of municipal fluoridation is so extremely low that any other means used to provide children with fluoridated water, to the exclusion of other age groups, would be most expensive.

It has been frequently stated that fluorides are toxic substances and consequently can provide our enemies with opportunities for sabotage. It is true that fluoride compounds are toxic at high concentrations, and so are almost all other substances. For instance, chlorine, a

[1] Chemistry of Fluorine, from "Our Children's Teeth: A Digest of Expert Opinion Based on Studies of the Use of Fluorides in Public Water Supplies," Committee to Protect Our Children's Teeth, Inc., New York, Mar. 6, 1957.

toxic gas used in World War I, is almost universally used today to purify our drinking water. Water itself consumed in large amounts is harmful. The small quantity of fluorides used for water supplies is far from being toxic. It has been computed that about 500 gal water fluoridated at 1.0 ppm would have to be consumed at one sitting to provide enough fluorides for a lethal dose. The toxic properties of fluorides in massive amounts have led to the suggestion that through an accident or sabotage dangerous levels of fluorides might be introduced into water at the treatment plant. In order to provide a lethal dose, however, about 40 tons of sodium fluoride would have to be provided for each million gallons of water. This is practically impossible because the chemical feeding device would be designed to supply only about 19 lb per million gal and no more than a month's supply (600 lb) would ordinarily be on hand.

PROBLEMS ARISING IN THE WATER PLANT BECAUSE OF FLUORIDATION

One of the most frequently heard objections to fluoridation is the unfounded assertion that fluorides cannot be fed accurately into the water. While it is true that in a few instances the level of fluorides has varied widely, this does not mean a good job cannot be done or that most communities and their waterworks are not maintaining the fluoride level within narrow limits. A summary of several years' sampling of the fluoridated water at Grand Rapids, Michigan, and Newburgh, New York (Figs. 8.18 and 8.19), indicates that the vast majority of samples can be maintained within 0.1 ppm of the desired amount. Experiences in other places[2] are similar. As indicated in Chap. 8, the feeders used for fluoride compounds are very accurate and water-plant operators are trained to maintain their feeders to preserve this accuracy.

Three possible causes of fluoride variations in the water after it leaves the treatment plant have been mentioned as being undesirable. One involves the discovery of a fluoride level in the distribution system lower than at the water plant. This has been found to occur only during the early stages of fluoridation. It might be caused by either dilution of the fluoridated water with the unfluoridated water in the mains and reservoirs or by absorption by compounds on the lining of

[2] "Water Fluoridation Practices in Major Cities of the United States," New York University College of Engineering, 1959.

the mains. It has been known for many years, for instance, that the tubercles forming on the inner surfaces of mains accumulate varying amounts of negative ions from the water. These include sulfates, chlorides, and fluorides. However, contrary to the opinion of some critics of fluoridation, these do not suddenly dissolve or become dislodged and thereby increase the fluoride level in the water. As observed many times, "the tubercles adhered tightly to the cast-iron pipe, and it was necessary to scrape the pipe forcibly with a metal tool to collect the sample for analysis. Hence, it is believed that the encrustment would not be removed by turbulent water flow."[3]

Hazards to Operators

The possible dangers in handling fluoride compounds and the precautions and techniques practiced are described in Chap. 12. It is perhaps most significant that only one waterworks employee has ever been known to have been harmed because of handling fluorides, and this occurred because of his failure to observe ordinary safety precautions. In every detailed survey undertaken to discover any harmful effects, the degree of exposure has been found to be many times below the accepted safe level.[4]

Effects Caused by Corrosion

Many people claim that fluoridation makes water corrosive, perhaps because they confuse treating water with chlorine (which is very corrosive) and treating it with fluorides. The American Water Works Association[5] has stated: "It may be stated categorically that the modification of chemical characteristics in the public water supply, treated with 1 to 1.25 parts per million of fluoride containing materials, does not increase or decrease the corrosivity of the water."

COMMERCIAL AND INDUSTRIAL USE OF FLUORIDATED WATER

There are very few references relating to the effects of fluoridated water on industrial processes because very few such problems have

[3] Experiences in Applying Fluorides, *J. Am. Water Works Assoc.*, **49**(10): 1239 (1957).
[4] *Ibid.*
[5] Harry E. Jordan, "Commercial and Industrial Use of Fluoridated Water," letter to AWWA Membership, Feb. 9, 1956.

Technical Objections (Engineering, Chemical, Industrial, Economic)

occurred. A statement published by the Baltimore, Maryland, Association of Commerce explains the reason for this:

> We have discussed this matter with a representative cross-section of industries which use city water in their processes. Sixteen major companies were selected, with diversified products in the chemical, porcelain enamel and frit, drugs, soap, porcelain insulators, food, beverages, and plating industries, and including large users of water in steam generation. These companies were asked if the proposed addition of fluoride to city water would create difficulties in their processes, products, or equipment. All of the companies have now reported that the addition of fluoride to city water would have no adverse effect on their operations.

This is typical of many such statements relating to the experience of many diverse industries using fluoridated water.

There have been, in fact, only two adverse effects on any industrial processes. A manufacturer at Charlotte, North Carolina, found an increase in the number of shattered blocks of manufactured ice, but this was corrected by the addition of a small amount of ammonium chloride to the water prior to freezing. The other effects related to the increase in fluoride which might occur if manufactured baby foods were prepared by processes resulting in concentrating fluoridated water by evaporation. The possibility of any criticism of the products made by the baby-food manufacturer using this process has been avoided by using water which has been defluoridated. In this regard the Federal Food and Drug Administration has ruled:

> ... commercially prepared foods within the jurisdiction of the Act, in which a fluoridated water supply has been used in the processing operation, will not be regarded as actionable under the Federal law because of the fluorine content of the water so used, unless the process involves a significant concentration of fluorine from the water. In the latter instance the facts with respect to the particular case will be controlling (July 17, 1952).

References relating to specific industries or products are:

Baking (bread)—Special Bulletin No. 66, July 6, 1950, American Institute of Baking, Chicago, Illinois.

> The addition of fluoride ion in concentrations up to 10.0 ppm in sponge and dough water has no effect upon bread quality. Bakers in a community that plans to encourage fluorine in a city water supply

as part of a program to reduce the incidence of dental caries should anticipate no difficulties in using such water for bread production.

Beer (breweries, yeast activity, fermentation)—R. S. Slater, The Influence of Fluorine upon Fermentation, *Proc. Am. Soc. Brewing Chemists*, April, 1951.

> When yeast first encounters wort (unfermented malt) containing 1 p.p.m. F, its metabolism is stimulated slightly. 5 p.p.m. F has a similar effect. Depending upon the source of the fluorine, 10 p.p.m. may have a slightly stimulating or depressing action. At concentrations of 25 p.p.m. fluorine or more, the fluorides are definite inhibitors of yeast activity.

The fears expressed by the brewing industry in Milwaukee and St. Louis that fluoridated water would affect their products proved to be unfounded.

Canning (preserved foods)—Research Bulletin No. 6051, American Institute of Canning, Washington, D.C.

> The effects with which canners would be concerned, aside from physiological considerations, would be those on the flavor, color, or texture of the product and on corrosion of the container. The known properties of fluoride are not such as to suggest the likelihood of any such effects from the amounts involved here, and the lack of any reported experiences attributable to such effects, indicates that they would be very difficult, if not impossible, to demonstrate experimentally.

Sewage Treatment—Effects of Fluoride Concentration on Sludge Digestion, *Sewage and Ind. Wastes*, January, 1955.

> It appears unlikely from these results that the presence of a normal fluoride concentration would produce detrimental effects on the digestion of sewage solids.

Since these statements were published, no adverse effects in these or any other industrial or commercial processes have been reported.

OTHER METHODS FOR ADMINISTERING FLUORIDES

It has been shown that the ingestion of fluorides at the optimum levels is effective in reducing dental caries and is not in any way harmful. Many people who are convinced of these conclusions, how-

Technical Objections (Engineering, Chemical, Industrial, Economic)

ever, believe that some means other than the water supply should be used for providing the fluoride. If such other means could be found, then only that portion of the population which would be primarily benefited would receive it. The following alternatives have been most frequently suggested, and some of the reasons for their present impracticability as compared with water fluoridation are given.

Topical Applications

Many studies have been completed on the effects of applying strong fluoride solutions directly to the teeth of children and adults. Approximately a 40 per cent reduction in DMF rates among children has been reported. The principal objections to topical applications are:

1. It is not so effective as water fluoridation (40 per cent against 60 to 70 per cent reduction).
2. It is time-consuming. Four applications by a dentist or trained assistant at intervals of about a week should be completed at four age periods (3, 7, 10, and 13 years of age) with each application requiring about an hour.
3. The time necessary to complete the treatment makes it very expensive.
4. At present, there are not enough dentists available to do this for all children.
5. As a public health measure it would fail, principally because the individual effort required of the parents to complete the treatments.

Tooth Paste, Mouthwash, and Chewing Gum

These means for providing fluoride are thought to be similar to topical applications in their action but are self-administered. Tooth paste containing stannous fluoride has recently proved to be effective, but care must be exercised in using it because of possible excessive fluoride intake, particularly among children who might consume it for its candylike flavor. This danger would be even more pronounced in areas containing excessive natural fluorides in the water supply. Fluoride-containing mouthwashes and chewing gum have not proved effective. These means for providing optimum fluorides would also fail because of the laborious individual efforts required to make them effective.

Fluoride-containing Tablets

Tablets containing specific amounts of fluoride (generally 1.0 mg as the fluoride ion) have been used in two ways—taken daily as a pill or dissolved in the children's drinking water. At present, no results have been published showing conclusively that this method is effective. It would appear, however, that in order to obtain results comparable with those observed in areas containing naturally the optimum amounts, some consideration should be given to varying the amounts of fluoride administered, depending on the age of the child. This is illustrated in Chap. 6, in describing the varying amounts of water consumed by children as their age increases. Many projects have been started in schools, particularly in Switzerland, where there are relatively few public water supplies, but at school age the greater portion of permanent-tooth dentition is completed. Aside from excessive cost, the strongest objection to this method is the almost insurmountable difficulty of maintaining for at least eight years daily constancy, on the part of the parents, in providing the tablets. In addition, " . . . the daily consumption of tablets likewise raises questions of effectiveness and practicality. In the hands of trained personnel at the water treatment plant, fluoride levels can be precisely controlled. But experience with other home remedies—even the aspirin tablet—prompts caution."[6] The philosophy that "if one tablet is good, two are better" may produce harm. A child's accidental ingestion of a large number of tablets is recognized as a great hazard by those familiar with accidents in the home.

Bottled Water

The use of bottled fluoridated water is probably the best substitute for a fluoridated public supply because, of all the alternatives, fluoridated drinking water has been proved to be the most effective, least harmful, and least expensive. However, the cost of distributing bottled fluoridated water would be prohibitive, particularly if all children had to be served fluoridated water in this manner. For New York City the cost per person was estimated to be about $18 per year against a cost of 10 cents for fluoridating the public supply.[7] In addi-

[6] Report to the Mayor on Fluoridation for New York City Board of Health, City of New York, Oct. 24, 1955.
[7] Ibid.

Technical Objections (Engineering, Chemical, Industrial, Economic)

tion, bottled water is inconvenient, and to be effective it requires, on the part of the parents, the same continuous, conscientious effort as was described for tablets.

Salt and Bread

Objections to these media as vehicles for fluorides for children are:
1. There is wide variation in consumption of bread and salt by children.
2. No proof exists that fluorides would be effective if incorporated in salt or bread.
3. No data are available on what amounts of fluoride to use.

Milk

Milk has many times been seriously considered a likely alternative, probably because milk is such a universally used food for children. There are, however, such serious objections to fluoridating it for children's consumption that it cannot be attempted with our present knowledge. Some of these objections are:
1. It is not known how much fluoride should be added to milk. The amount obviously cannot be the same as for water. If it were, infants, who subsist largely on milk, would obtain too much fluoride.
2. The variations in children's consumption of commercially distributed milk are too great when such factors as other sources of milk are considered (for instance, dried or condensed milk).
3. No proof exists that fluorides in milk would be effective.
4. The cost would be very high. For school children alone in New York City the cost of fluoridated milk would amount to $2.14 per child as compared to 10 cents for fluoridated water.[8]

Considering the many objections to every alternative to fluoridated water, it can be seen that the effectiveness and safety of fluoride administered in the drinking water at the recommended levels are better established than by any other means.

[8] *Ibid.*

CHAPTER 6 *Fluoride Concentration Desired in the Treated Water Supply*

Most raw (untreated) waters contain fluorides. Surface waters—those sources comprising rivers, lakes, ponds, canals, creeks, and cisterns—generally do not exceed 0.3 ppm fluoride except when contaminated with industrial wastes or sewage. Industrial wastes, particularly from the steel, aluminum, and fertilizer industries, cause the receiving streams to vary widely in their fluoride content. This variation increases the difficulty of maintaining a constant fluoride level in water treated from such sources. If discharged into a stream from a community which is fluoridating its water supply, domestic sewage will raise the fluoride content of the receiving stream. Such sewage will contain the same fluoride concentration as the water supply, and its volume may constitute a considerable proportion of the stream into which it is discharged. Fluoride levels in surface sources vary according to the quantity of fluoride sources and of runoff; the higher the river stage, the lower the fluorides.

Ground waters (springs, wells, and infiltration galleries) are the sources used for most of the water supplies in this country. They also generally contain the higher fluoride concentrations (Fig. 1.1).

Fluorides usually occur in well waters, primarily because of the presence of fluorspar, phosphate rocks, or cryolite. The fluorides in such waters are present because they have dissolved and the fluoride ion is freed from one or more of these three sources. The fluoride deposits can be long distances from the point of appearance of the waters which have dissolved them. Many of the high-fluoride well waters in all the states extending from the Dakotas south to Texas

Fluoride Concentration Desired in the Treated Water Supply 49

have probably obtained their fluorides from deposits in the Rocky Mountains.

Fluorspar, a mineral containing fluorite or calcium fluoride, is generally found in veins underground. The largest deposits in this country are in limestone faults in southern Illinois and northwestern Kentucky. A suggested explanation for these deposits is that thermal waters containing fluorides ascended and spread into limestone crevasses and cavities. On cooling, the fluoride salts remained as solid material filling the voids. Such deposits are very widely found and may well be the source of fluoride in most well waters.

Tremendous quantities of fluorides are found in phosphate-rock deposits, but these are rather concentrated in large deposits in relatively few isolated areas. They generally occur as sedimentary deposits of marine origin. The presence of fluorine as part of the phosphate compound forming complex rock particles is responsible for the very low solubility of the rocks. This in part accounts for the large accumulations of such deposits in several widely spaced areas of the world. The waters found near phosphate deposits may contain fluorides, but the wells influenced by them are definitely limited in area.

Cryolite is found commercially only in Greenland. Other smaller deposits exist—one of the largest in the United States being on Pikes Peak. This may account for the fluoride concentration in the water supply serving Colorado Springs, Colorado, which contains 2.4 ppm.

Many well waters contain excessive fluorides (more than about 1.5 ppm), and in these cases the excess should be removed (see Chap. 16). Many thousands of such supplies, however, are deficient in fluorides, and the amount required can be economically and easily added. In most cases the natural fluorides in these sources vary but little, and for this reason the optimum fluoride level can be readily maintained with a minimum of equipment and supervision.

Table 6.1 shows the number of public water supplies in the United States, grouped according to the average fluoride content found in them. It is obvious that the vast majority are deficient in fluorides. The table also shows that approximately five hundred supplies should be either defluoridated or diluted to the point where the fluoride level is maintained at the optimum. Figure 1.1 shows the location of communities in which the water naturally contains 0.7 ppm or more of fluoride. The large number of supplies containing some, but not enough, fluoride raises the question of the practicability of supplementing the natural fluorides with enough to bring the concentration up to the optimum level. Usually it can be shown that even though very little more is needed, it is economically justifiable to add it.

TABLE 6.1. COMMUNITIES GROUPED ACCORDING TO FLUORIDE CONCENTRATIONS OCCURRING NATURALLY

Fluoride content	Number of communities	Population
0.0–0.6	16,570	93,098,000
0.7–1.1	896	4,254,728
1.2–1.4	266	602,323
1.5–2.9	454	975,750
3.0–4.9	131	399,687
5.0 and over	32	22,497
Total............	18,349	99,354,985

For example, if it costs $5.50 in a particular city to restore a DMF tooth, the savings realized in a community of 35,000 people (containing four hundred and ninety 13-year-old children) would amount to $5,390 among this age group alone, if the fluoride level were raised from 0.5 to 1.0 ppm fluoride.

Figure 6.1, based on DMF results of 57 studies in various cities, shows that an increase in fluoride level from 0.5 to 1.0 ppm produces a decrease of approximately two DMF teeth; and an increase of from 0.7 to 1.0 ppm produces a decrease of about one DMF tooth per person.[1]

A figure for the fluoride concentration of raw untreated water, for purposes of designing the fluoride installation, can generally be obtained from the local or state health department laboratories. The longer the record of such examinations, the more reliable the figure. In some areas, however, no fluoride determinations have ever been made. In these cases, the sampling should begin as soon as possible and conditions (river stage, season, period of industrial waste discharge, and the like) chosen to provide the minimum fluoride level. The maximum probable fluoride feeding rate can then be determined by subtracting the value for the minimum fluoride level found under these conditions from the value for the optimum fluoride level.

[1] David F. Striffler, Criteria to Consider When Supplementing Fluoride-bearing Water, *Am. J. Public Health,* **48:** 29–37 (1958).

Fluoride Concentration Desired in the Treated Water Supply

The optimum fluoride level in water (1.0 ppm in the mid-United States, in the latitude of Kansas City–Chicago) has been shown to be that level which produces the greatest protection against caries with the least hazard of fluorosis.

In Chap. 2 it was shown that fluoridation is most advantageous when ingestion of the optimum quantities of fluorides commences at

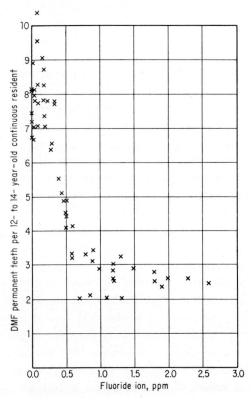

FIG. 6.1. Relationship between fluoride levels and prevalence of dental decay.

birth and continues essentially uninterrupted. Substantial benefits are also realized if the starting period is somewhat delayed, but with each successive year's absence of fluoride consumption there is corresponding increase in dental decay. No benefits to adults (if they had not previously consumed fluoridated water) have been demonstrated so far.

In Chap. 2, it was brought out that a *remarkably* constant quantity of fluoride appears to be ingested by children living in the same environment. This was concluded by observing the effects on the part of the body most sensitive to fluoride; viz., the teeth. When the optimum fluoridated water is consumed, starting at birth, no disfiguring fluorosis has ever been observed and the best reductions in tooth decay have been realized. This does not mean, however, that children of all ages consume the same amounts of fluoride every day. The very youngest children get the smallest quantities, their diet being largely milk and cereal. As they grow older, however, the amount of water consumed, together with the quantities of fluoride-containing foods eaten, increases.

The concentration of 1.0 ppm was considered the optimum after direct observations of thousands of children's teeth. It was not based on any direct or accurate knowledge of how much water children drink at various times and at different places. Neither was this level established on the basis of measuring out to them a specific amount of fluorides in their daily diet. Instead, these thousands of children obtained almost an optimum quantity of fluorides solely by means of a natural selection of foods and drinking water. Inasmuch as it has been shown that the quantity of fluorides in the foods consumed by children changes very little, the variations in dental effects among children were brought about almost entirely by the differences in fluorides in their drinking water.

According to the study of the "21 cities" described in Chap. 1, when the drinking water contains the optimum quantity of fluoride, then the fluorides consumed are obtained almost entirely from the water. A small proportion is obtained from foods. Almost all foods contain some fluorides. This is not in the least remarkable in view of the vast quantities and widespread occurrence of the compounds of fluorine found in nature. A number of the more commonly consumed foods and beverages and their fluoride contents are listed in Table 6.2.

Most vegetables and meat contain less than 1.0 ppm fluoride on a dry basis. The outstanding exceptions are tea, which may contain as much as 60 ppm fluoride, and seafoods, which contain up to about 30 ppm. These foods are not likely to be an important part of children's diets. One notable exception to this is the almost exclusive diet of fish used by the people of Tristan da Cunha, an isolated island in the middle of the South Atlantic Ocean due west of Cape Town. It has been found that their fish diet contributed as much fluoride (with an accompanying absence of dental decay) as if the drinking

TABLE 6.2. FLUORINE CONTENT OF FOODS AS REPORTED
IN THE LITERATURE†

Food	Fluorine, ppm	Food	Fluorine, ppm
Fluorine reported in food as consumed			
Milk	0.07–0.22	Pork chop	1.00
Egg white	0.00–0.60	Frankfurters	1.70
Egg yolk	0.40–2.00	Round steak	1.30
Butter	1.50	Oysters	1.50
Cheese	1.60	Herring (smoked)	3.50
Beef	0.20	Canned shrimp	4.40
Liver	1.50–1.60	Canned sardines	7.30–12.50
Veal	0.20	Canned salmon	8.50–9.00
Mutton	0.20	Fresh fish	1.60–7.00
Chicken	1.40	Canned mackerel	26.89
Pork	0.20		
Fluorine reported in dry substance of food			
Rice	1.00	Honey	1.00
Corn	1.00	Cocoa	0.50–2.00
Corn (canned)	0.20	Milk chocolate	0.50–2.00
Oats	1.30	Chocolate (plain)	0.50
Crushed oats	0.20	Tea (various brands)	30.00–60.00
Dried beans	0.20	Cabbage	0.31–0.50
Whole buckwheat	1.70	Lettuce	0.60–0.80
Wheat bran	1.00	Spinach	1.00
Whole-wheat flour	1.30	Tomatoes	0.60–0.90
Biscuit flour	0.00	Turnips	0.20
Flour	1.10–1.20	Carrots	0.20
White bread	1.00	Potatoes (white)	0.20
Ginger biscuits	2.00	Potatoes (sweet)	0.20
Rye bread	5.30	Apples	0.80
Gelatin	0.00	Pineapple (canned)	0.00
Dextrose	0.50	Oranges	0.22

† F. J. McClure, Ingestion of Fluoride and Dental Caries. Quantitative Relations Based on Food and Water Requirements of Children One to Twelve Years Old, *Am. J. Diseases Children*, **66**(4): 362–369 (1943).

water contained as much as 1.0 ppm fluoride. Actually the drinking water had less than 0.2 ppm fluoride.[2]

In this country, many studies have been made to establish the importance of food as a contributor of fluorides. It has been shown that

[2] Reidar F. Sognnaes, Relative Merits of Various Fluoridation Vehicles, in James H. Shaw (ed.), "Fluoridation as a Public Health Measure," American Association for the Advancement of Science, Washington, 1954, pp. 179–192.

the average diet seldom contains more than about 0.3 ppm fluoride. This quantity is remarkably constant; food from various parts of the country shows very little variation in fluoride levels. The fluoride content of plants appears to depend on the species and not primarily on the type of soil; an excess of fluorides in the soil in which the plants are grown does not increase the fluoride content of the plants. The same observation was made regarding the fluoride content of cow's milk. Milk has generally been found to contain less than about 0.2 ppm fluoride. When cows consume water containing even as much as 8.0 ppm fluoride, their milk has never been found to exceed 0.3 ppm fluoride.

Children up to 12 years of age consume between 1,200 and 2,500 ml water per day, the variation depending on their age (1 to 3 years —1,200 ml; 10 to 12 years—2,500 ml) and environment.[3] If this water contained 1.0 ppm fluoride, then from this source alone between 0.4 and 1.1 mg fluoride would be obtained. The maximum variation of fluoride from food and water for this age group would be as shown in Table 6.3.

TABLE 6.3. SUMMARY OF ESTIMATED DAILY INTAKE OF FLUORIDE FROM FOOD AND FROM DRINKING WATER (DRINKING WATER CONTAINING 1 PPM FLUORIDE AND DRY SUBSTANCE OF FOOD CONTAINING 0.1 TO 1 PPM FLUORIDE)

Age, years	Body weight, kg	Daily fluoride intake			
		From drinking water, mg	From food, mg	Total from food and drinking water, mg	Total as mg/kg of body weight
1–3	8–16	0.390–0.560	0.027–0.265	0.417–0.825	0.026–0.103
4–6	13–24	0.520–0.745	0.036–0.360	0.556–1.105	0.023–0.085
7–9	16–35	0.650–0.930	0.045–0.450	0.695–1.380	0.020–0.068
10–12	25–54	0.810–1.165	0.056–0.560	0.866–1.725	0.016–0.069

With the establishment of the relatively minor contribution of foods as a source of fluorides for children, it was concluded that the variations in the dental effects of fluorides were caused entirely by the differences in fluoride concentrations of the public water supplies.

[3] Optimum Fluoride Intake for the Prevention of Dental Caries, National Academy of Sciences, National Research Council, Publication 294, November, 1953, p. 7.

As a result of Dr. H. Trendley Dean's studies on this subject during the 1930s, it was possible for his colleague, Dr. F. A. Arnold, to conclude during 1943:[4]

> The results of both epidemiologic, chemical, and experimental studies suggest that the addition of small amounts of fluoride, not to exceed 1 part per million, to fluoride-free public water supplies may be a practical and efficient method of markedly inhibiting dental caries in large population groups.

He also observed that:[5]

> the continuous use throughout the formative period of the tooth of water containing about 1 part per million of fluorine will result in an incidence of approximately 10 per cent of the mildest form of dental fluorosis.

As early as 1939 Dr. Gerald J. Cox,[6] then with the Mellon Institute, wrote:

> Regulation of fluorine should be directed at an optimum intake of the element. In particular, the fluorine content of the water supply can be varied seasonally to compensate for varying water consumption. Climatic differences will make it necessary for each locality to find its own standards for addition of fluoride to the water supply.

In clarifying this conclusion, Dr. F. A. Arnold[7] stated in 1943:

> It cannot be too strongly stressed that the fluorine-dental caries theory rests basically upon the question of the amount of fluoride ingested by the population. For example, a water containing 0.5–0.7 parts per million of fluoride in southwestern United States may have an effect equal to that of a water of 1.0–1.5 parts per million of fluoride in the north central section of this country. For this reason, it may be desirable to base precise estimates upon the more acceptable biologic measurement of fluoride intake on a population, the index of dental fluorosis.

Biological measurements were in fact used later several times to determine the optimum fluoride levels for areas where water con-

---

[4] F. A. Arnold, Role of Fluorides in Preventive Dentistry, *J. Am. Dental Assoc.*, **30**: 499–508 (1943).

[5] H. Trendley Dean and F. A. Arnold, Endemic Dental Fluorosis or Mottled Enamel, **30**: 1278 (1943).

[6] Gerald J. Cox, New Knowledge of Fluorine in Relation to Dental Caries, *J. Am. Water Works Assoc.*, **31**: 1926 (1939).

[7] *Op. cit.*

sumption, because of climatic factors, varied from that in the north central area of the country. Dr. Dean probably did the earliest work of this sort when he attempted to determine the optimum fluoride levels for Georgia and Florida. His results, though very sparse, were published during 1950.[8] He found that where the mean annual temperature is about 68°F, mild dental fluorosis may be as prevalent in areas using water containing 0.7 ppm fluoride as in cooler areas (mean annual temperature 49°F) where 1.0 ppm fluoride waters are used.

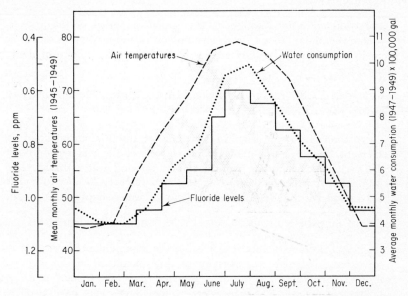

FIG. 6.2. Variations of fluoride levels with temperature at Charlotte, North Carolina.

Based on these observations, the fluoridation project at Charlotte, North Carolina, was designed to provide a more constant fluoride intake by varying the fluoride levels seasonally. Dr. Zachary Stadt (with Mr. Robert S. Phillips, the water-plant superintendent, who designed the system) estimated seasonal variations in fluid intake by determining the amount of carbonated beverages sold in Charlotte. From these figures a graph (Fig. 6.2) was constructed. The fluoride levels maintained give a weighted average of 0.9 ppm during the year.

The objective of this procedure is to provide a more uniform fluo-

[8] F. J. Maier, Fluoridation of Public Water Supplies, *J. Am. Water Works Assoc.*, **42**(12): 1120–1132 (1950).

Fluoride Concentration Desired in the Treated Water Supply

ride intake throughout the year regardless of temperature and water consumption. When such water consumption is lowest (in the winter), the fluoride level in the water is the greatest; in the summer, when water consumption rises, the fluoride concentration in the water is decreased. This method of fluoridating water may permit a higher fluoride level to be maintained (with a correspondingly greater DMF reduction) with no increase in fluorosis. This idea appears to have considerable merit, although the procedure is no longer practiced at Charlotte.

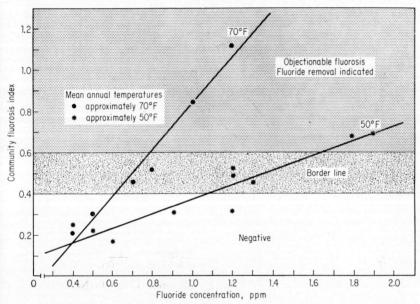

FIG. 6.3. Relationship between fluoride levels, fluorosis, and temperature. (*USPHS*.)

Later studies in California and Arizona, where temperatures are considerably above the average of other parts of the United States, showed a definitely lower optimum fluoride level. This was demonstrated by observing dental-fluorosis prevalence in places with various natural fluoride levels in their public water supplies and also by estimating the actual quantities of water ingested by children of various age groups and weights. In areas where the mean annual temperatures are greater than 70°F, the optimum fluoride concentration should not exceed 0.8 ppm. This relationship, compared with Dr. Dean's data from areas with mean annual temperatures of 50°F, is shown in Fig. 6.3.[9] Later a formula was derived for determining fluoride levels at any temperature:[10]

$$\text{ppm fluoride} = 0.34/E \qquad (6.1)$$

where E is the estimated average daily water consumption for children through 10 years of age in terms of ounces of water per pound of body weight. E is obtained from the formula:

$$E = -0.038 + 0.0062 \times \text{avg max temp, }°F \qquad (6.2)$$

For example, it is desired to determine the optimum fluoride level to maintain at a place having an average maximum temperature (the mean of all the daily maximum temperature readings observed during the period considered, based if possible on observations obtained for at least five years) of 55°F. Then:

$$E = -0.038 + 0.0062 \times T°F$$
$$E = -0.038 + 0.0062 \times 55 = -0.038 + 0.342$$
$$E = -0.304$$

Then ppm fluoride [from formula (6.1) above] $= 0.34/0.304 = 1.1$ ppm.

Table 6.4 shows optimum fluoride levels suggested by the authors

TABLE 6.4. MEAN MAXIMUM TEMPERATURES AND CORRESPONDING RECOMMENDED OPTIMUM FLUORIDE CONCENTRATIONS

Mean maximum temperatures	Recommended optimum fluoride concentration
50.0–53.7	1.2
53.8–58.3	1.1
58.4–63.8	1.0
63.9–70.6	0.9
70.7–79.2	0.8
79.3–90.5	0.7

of the paper[11] in which this formula appears.

While these formulas and tables will provide a reasonable estimate of the optimum fluoride content to maintain, the final estimate should

[9] Donald J. Gallagan, Climate and Controlled Fluoridation, *J. Am. Dental Assoc.*, **47**: 159–170 (1953).
[10] Donald J. Gallagan and Jack R. Vermillion, Determining Optimum Fluoride Concentration, *Public Health Repts.* (*U.S.*), **72**(6): 491–493 (1957).
[11] *Ibid.*

Fluoride Concentration Desired in the Treated Water Supply

be checked with the state health department having jurisdiction. Many health departments have made careful studies of the optimum levels within their states and may have more accurate estimates based on the manner in which local conditions (diet, habits, etc.) may affect the estimates derived in the manner described above.

The fluoride levels estimated in this manner are intended for use where no variation in concentration is contemplated on a seasonal basis. This estimate is based on biological observations in areas where

TABLE 6.5. FLUORIDE CONCENTRATIONS IN TREATED WATER

State	Winter level, ppm	Summer level, ppm
West Virginia	0.8–1.3	0.6–1.1
Rhode Island	1.2	1.0
Alabama	1.0	0.7
Oklahoma	1.0	0.8
Tennessee	1.0	0.8
Washington	1.5	0.6–0.7

fluorides occur naturally in the public water supplies and little if any seasonal variations in fluoride levels are experienced. We know that where the levels are constant, a certain reduction in dental decay occurs and no aesthetically noticeable fluorosis appears. As at Charlotte, however, it was thought that a greater degree of caries protection might be realized if the fluoride levels were changed as the consumption of water varied because of temperature or other environmental factors.

Some states suggest that the fluoride level be decreased during the warmer months and increased during the winter. Table 6.5 shows a summary of some of these recommendations.

CHAPTER 7 *Fluoride Compounds*
(Characteristics, Sources, Costs)

In Chap. 1 it was explained that the most commonly found minerals containing fluorides are fluorspar (which contains calcium fluoride), cryolite (which contains fluoride combined with aluminum and sodium), and apatite (which usually is a calcium compound of fluorides, carbonates, and sulfates). Fluorine as a chemical element (not combined with any other element to form a chemical compound) does not exist free in nature. Elemental fluorine is now being manufactured and is available in large quantities as a gas, but its cost and the extreme hazards associated with its handling will at present make it unsuitable as a water-fluoridating agent.

As fluoride compounds are estimated to constitute about 0.08 per cent of the earth's crust (with sea water usually containing about 1.0 ppm fluoride), fluorine ranks about thirteenth among the elements in order of abundance. Many of the fluorides found naturally in waters are probably derived principally from the three minerals mentioned above.

After a rain, water percolates down through the soil and becomes what is called "ground water." Some of this water later reappears in the form of springs or from wells and galleries. In passing through the earth, the ground water dissolves in varying amounts the minerals with which it comes in contact. In this manner we get waters of widely varying qualities. We find waters, for instance, which are called "hard" or "sulfurous" or "acidy" or "bitter." These terms are descriptive of the type of minerals dissolved or gases entrained in the water. Fluorides are also found naturally in many waters. These are dissolved from fluoride-containing minerals occurring in the passages of the water as it flows through the earth. Fluorides are found in almost all water supplies; some sources contain too much, some too little, and a few have just the right amount. The map (Fig. 1.1) shows how these supplies are distributed in the United States.

FLUORSPAR

Fluorspar is the principal source of the commercial fluoride compounds available at the present time, even though known deposits of phosphate rocks (apatites) may contain more fluorides. Technical and economic problems still remain to be overcome if fluorides obtained from apatites are to compete successfully with those formed from fluorspar. Cryolites have a higher fluoride content, but deposits are relatively very small.

Fluorspar is a mineral containing varying amounts of calcium fluoride. Pure calcium fluoride contains 51.1 per cent calcium and 48.9 per cent fluoride. Fluorspar as mined may contain as little as 30 per cent calcium fluoride, but as such this low quality cannot be used commercially without some purification or beneficiation. Commercial grades of fluorspar usually contain from 85 to over 98 per cent calcium fluoride.

Pure fluorspar is a lustrous, glasslike material almost always translucent or transparent. It may be colorless or range in color from blue to violet, amethyst, purple, green, red, or yellow. Deposits often occur in masses of very pure crystalline material mixed with pieces of galena, quartz, calcite, barite, sphalerite, and many other contaminants. Deposits occur in every part of the world, with substantial commercial production found in Argentina, Australia, Canada, China, Korea, England, France, Germany, Italy, Mexico, Newfoundland, Russia, Spain, Tunisia, the Union of South Africa, and the United States. The largest deposits so far found occur in the United States; in southern Illinois and northeastern Kentucky an area comprising 700 sq miles contains an estimated reserve of fluorspar sufficient to supply our needs for the next twenty years at the present rate of consumption. Smaller deposits are found in some of the Western states, Colorado, Nevada, Montana, and New Mexico being at present those producing the larger quantities.

Discoveries of fluorspar in this country occurred in New Jersey and other Eastern states during the period 1814 to 1816; the deposits in

Illinois were discovered a few years later. The first recorded use of fluorspar in the United States occurred in 1823, when 2 oz from Shawneetown, Illinois, were mixed with 4 oz sulfuric acid to make hydrofluoric acid. The value of fluorspar as a flux was known in 1529, but it was not until 1837 that some fluorspar mined near Trumbull, Connecticut, was used for this purpose in the United States in the smelting of copper ores. Up to about 1887 the principal use of fluorspar was in the manufacture of glass, enamels, and hydrofluoric acid, with a relatively minor amount used in the iron foundries and in the smelting of various ores. With its value as a flux established (for making the slag more fluid and as an aid in removing impurities) in the production of open-hearth and electric-furnace steel, consumption rapidly increased. Imports and domestic production rose to over 300,000 tons per year prior to World War II. Normal consumption has averaged over 500,000 tons per year since then. For example, during 1961 shipments from domestic mines amounted to over 205,000 tons while imports were more than 505,000 tons.

Fluorspar ore is generally mixed with many other rocks, which must be separated from it before it is commercially acceptable. The methods of treatment depend on the quality of the ore, the kind of impurities which must be removed, and the ultimate use to be made of the purified fluorspar. There are some few deposits which contain such pure fluorspar that merely removing the impurities by hand produces a satisfactory material. In most cases, however, the ore must be subjected to elaborate separation techniques, which include washing, screening, gravity separation by jigs and tables, and flotation. In addition to the usual impurities, such as calcite (calcium carbonate), quartz (silicon dioxide), and clay and sand (which are not harmful except that they dilute the ore), there are also marketable impurities, including barite (barium sulfate), galena (lead sulfide), sphalerite (zinc sulfide), pyrite (iron sulfide), and many other compounds of lead, zinc, and iron. In many cases the sulfides are in such quantity that they can be economically recovered.

Fluorspar must be low in silica and most other impurities for the manufacture of hydrofluoric acid and other fluoride compounds. This degree of purity can be obtained only by a special purification process, which usually includes flotation. A typical process would be:

The ore as it comes from the mines is fed through a crushing system to reduce the size to about $3/8$ to $1\frac{1}{2}$ in. A ball mill then further reduces this size to a powder of a size in the order of 35 to 200 mesh.

Most ores in the United States require removal of zinc and lead sulfides, which is done by means of the flotation process. This process requires the flotation of the impurities from the fluorspar by means of chemicals added to the fluorspar slurry. Soda ash is added to hold the pH at 8.5 to 10. Other reagents usually include a frother, oleic acid (for collector purposes), and quebracho in order to depress calcite. The mixture is passed from three to as many as seven times through flotation tanks until the desired purity of fluorspar is reached. The final flotation concentrate is pumped to a thickener from which the pulp (containing 50 to 60 per cent solids) is pumped to a drum filter. The filter cake goes to a rotary dryer, which reduces the moisture to about 1 per cent. The material is then ready for bagging and shipment.

Fluorspar is now used in water fluoridation in two ways: the mineral is dissolved at the water plant and fed as a liquid into the water, or it provides the principal raw material in the manufacture of various fluoride-containing compounds which are available commercially and can be fed either as a liquid or solid directly into the water to be treated. The primary compound manufactured from fluorspar is hydrofluoric acid, which is used in the preparation of most fluoride-containing salts. This acid is obtained by the acidulation of fluorspar using sulfuric acid:

$$CaF_2 + H_2SO_4 \rightarrow CaSO_4\downarrow + 2HF$$

Hydrofluoric acid could be used (and was used in one municipal water plant) directly for fluoridating water, but it is generally too difficult to handle in water plants. Instead the acid is used to make a large variety of compounds which can be used for fluoridation. These include sodium fluoride.

The first process (dissolving the fluorspar at the water plant), however, is by far the most economical in that fluorspar is less expensive than any other available fluoride compound. The advantage of fluorspar in this regard can be realized by comparing the cost of its available fluoride ion with that of the commonly used fluoridation compounds shown in Table 7.3. Sodium silicofluoride is three times as expensive; ammonium silicofluoride is three and one-half times as expensive; sodium fluoride is six and one-half times as expensive; magnesium silicofluoride is seven times as expensive; and hydrofluosilicic acid is nine times as expensive.

It is for this reason that the fluorspar-dissolving process should be used whenever other factors in a particular water plant permit the use of this inexpensive material. The process of using fluorspar involves a means for dissolving calcium fluoride, which is almost insoluble in water. It has long been known, however, that calcium fluoride is soluble in strong acids or in solutions of aluminum compounds. If strong acids were used to dissolve fluorspar, the resulting mixture would be more expensive to prepare than many of the other compounds used for fluoridation. The use of solutions containing aluminum compounds is considerably more practical in that one of the more commonly used chemicals routinely employed in water plants as a coagulant of impurities is filter alum (aluminum sulfate).

On the basis of this chemical phenomenon, it was found[1] that by varying the concentration of alum in solution, any desired quantity of fluoride could be obtained from fluorspar. A saturated fluoride solution contains about one-tenth of the alum concentration. For example, a 10 per cent alum solution contains 1 per cent fluoride; a 30 per cent alum solution produces a saturated fluoride solution containing 3 per cent fluoride (30,000 ppm fluoride ion). Any other fluoride strength between these extremes can be obtained by changing the strength of the alum solution.

Before a saturated fluorspar solution in alum can be obtained, the following conditions must be satisfied:

1. Sufficient time for dissolving must be provided—usually about two hours at ordinary room temperatures (above about 60°F).

2. An excess of fluorspar must always be present in the alum solution.

3. Vigorous stirring must be continued during the dissolving period.

All these conditions can be met by an appropriate design of the dissolving tank. The first such tank, installed at the Bel Air, Maryland, water plant in July, 1956, is 2 ft in diameter and 4½ ft high. Its capacity is about 75 gal. As shown in Fig. 7.1, a circular weir around the inside of the tank provides a means for collecting and removing the liquid comprising the dissolved fluoride in the alum solution, which rises through the bed of fluorspar in the bottom of the tank. A central tube of stainless-steel pipe is fixed in the middle of the tank to support the mixer, and a cone is attached to the bottom of the pipe.

The device is placed near the source of alum solution, as shown in

[1] F. J. Maier and E. Bellack, Fluorspar for Fluoridation, *J. Am. Water Works Assoc.*, **49**(1): 34 (1957).

Fluoride Compounds (Characteristics, Sources, Costs) 65

Fig. 7.2. A small solution feeder previously used for feeding alum solution into the raw water is arranged so that it discharges into the central tube of the dissolver. The dissolver is charged with about 300 lb fluorspar, and the alum solution is made up by dissolving about 50

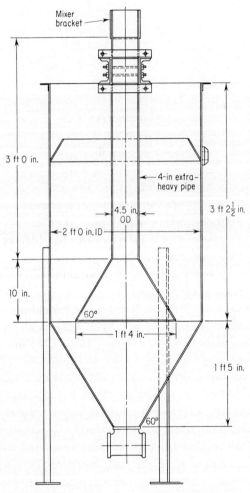

Fig. 7.1. Fluorspar dissolver. (*USPHS.*)

lb alum in 50 gal water (forming an alum solution of about 11 per cent strength). By adjusting the alum feeding rate, 1.0 ppm fluoride can be fed into the raw water along with 10 ppm alum. With these ratios, this dissolver can fluoridate up to 6 million gal water for each

Fig. 7.2. Fluorspar-dissolving equipment at Bel Air, Maryland. (*USPHS.*)

100 lb fluorspar. After this amount of water is treated, an additional 100 lb fluorspar is added to the tank to ensure a constant fluoride level and provide an excess. A clear effluent from a dissolver of this size can be obtained at upflow rates of 555 ml per min. This is equivalent to treating 2.3 million gal (1,600 gpm) water per day with 1.0 ppm fluoride. This rate can be increased to 666 ml per min, sufficient to treat 2.8 million gal (1,940 gpm) water per day at 1.0 ppm fluoride, but the effluent from the tank is slightly cloudy. The cloudiness of the solution is caused by calcium sulfate (gypsum) particles formed by the reaction of calcium fluoride and aluminum sulfate. At the lower rate (below 555 ml per min) no visible gypsum particles are suspended.

Calcium sulfate is produced at the rate of 1.7 lb per lb fluorspar dissolved. This must be removed from the tank whenever more fluorspar is added, for otherwise the tank would fill up with solids.

Fluoride Compounds (Characteristics, Sources, Costs)

At Bel Air this removal is accomplished by siphoning off a slurry of gypsum until enough room is provided in the tank for 100 lb fluorspar. A newer design (for the town of Rosiclare, Illinois) provides automatic removal of the gypsum by controlling the rate of alum addition. The flow of alum solution into the dissolver is regulated so that

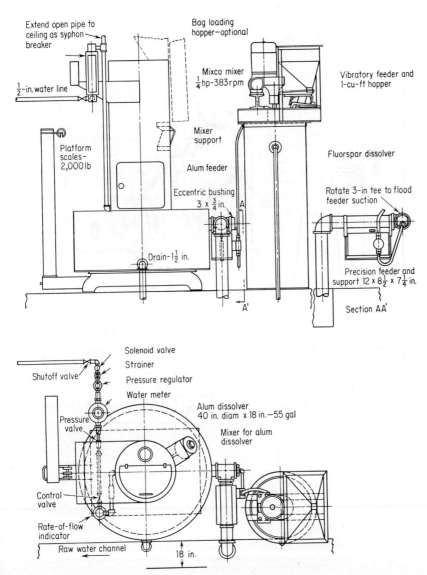

Fig. 7.3. Alum and fluoride systems at Rosiclare, Illinois. (*USPHS.*)

the gypsum is carried upward with the alum-fluoride mixture and discharged into the raw water. The presence of gypsum at this point is in no way harmful; gypsum particles in fact provide desirable nuclei for the formation of floc when the raw water is relatively clear. Drawings of the Rosiclare installation (Figs. 7.3 and 7.4) show the pro-

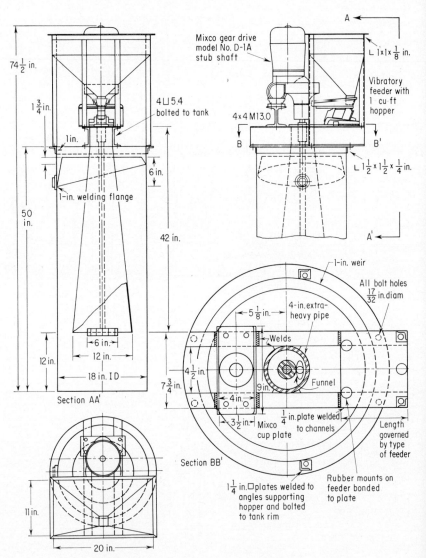

Fig. 7.4. Fluorspar dissolver at Rosiclare, Illinois. (*USPHS.*)

cedure for changing the fluoride and alum doses independently and for continuously replenishing the fluorspar in the dissolving tank. A constant rate of alum-solution addition is provided in the dissolver to suspend the proper amount of gypsum. The fluoride dose is changed by varying the alum concentration to the dissolver (changing the alum concentration is done by changing the alum-water ratio from the alum-feeder dissolving chamber). The alum dose is changed by varying the settings on the alum feeder. The alum-solution concentration is controlled by supplying the correct amount of water into the alum dissolver.

The cost of fluorspar dissolvers varies widely, depending primarily on their size and variety of appurtenances. Such costs have in all cases been written off within a year or two because of the marked savings realized through the use of the less expensive fluorspar.

Fluorspar is available commercially in (1) various grades of purity, (2) various degrees of flotation, and (3) different particle sizes. Any grade or type can be used in the dissolvers described here. Various local conditions, however, may indicate the desirability of a particular grade.

Broadly, there are three grades of purity. The metallurgical grade is used primarily for steelmaking and contributes 55 per cent of the total fluorspar demand. It should contain about 85 per cent fluorite (CaF_2) and less than 5 per cent silica. This is the cheapest grade and costs about $43 per ton. Similar imported material costs $23 per ton.

The ceramic grade is used in the manufacture of glass articles, clay products, and enamels for metal coatings. It is usually at least 95 per cent fluorite and contains less than 2.5 per cent silica and less than 1.5 per cent calcium carbonate. It costs about $48 per ton.

Acid-grade fluorspar is the best material available from processing plants and constitutes about one-third of the total quantity of fluorspar consumed. This grade is used both for the manufacture of hydrofluoric acid (and from this acid many fluoride salts are made) and also as the principal ingredient of synthetic cryolite. (The use of cryolite is described on page 87.) This grade should be at least 97 per cent fluorite and contain between 1.0 and 1.5 per cent silica. The cost of this grade is about $53 per ton.

Fluorspar generally contains flotation reagents after processing. At the water plant these prevent the fluorspar from sinking immediately into the alum solution. In smaller water plants, where batch operations are practiced, this may cause the fluorspar to float on the surface of the alum solution and be carried away undissolved in the foam formed by the flotation reagents. To prevent this, some processors can provide material from which the flotation reagents have been removed. This is done by raising the drying temperature (the last step in the process described on page 63) to about 400°F. Inasmuch as this does not involve an extra step in processing, the extra cost, if any, of this processed material is nominal. Some foreign deposits of fluorspar, still relatively undepleted, are of sufficient purity to make the flotation process unnecessary. The product imported into this country from such deposits contains no flotation reagents and consequently does not require heating or other treatment to remove them or minimize their effects.

Fluorspar is available in many different sizes, according to the demands of the consumer. Metallurgical grades, for instance, can be obtained as a gravel, in lumps, as artificial pellets, or as flotation concentrates of various sizes. For fluoridation, the finer the material, the more readily it dissolves. If too fine a material is specified, however (for instance, if 75 per cent of the material is finer than 325 mesh), the cost increases rapidly because of the special handling required in obtaining the finer portion. In general, the flotation concentrates as they come from the drying ovens are satisfactory if the fineness is as good as that available for most acid-grade materials.

Typical screen analyses of two common grades are:

	Acid grade, per cent	Ceramic grade, per cent
On 100 mesh	17	6
100–200 mesh	30	8
200–325 mesh	31	28
Through 325 mesh	22	58

Fluorspar is available from the following companies:

Ceramic Color & Chemical Company, New Brighton, Pennsylvania
Continental Ore Corporation, New York, New York
Eagle Fluorspar Company, Salem, Kentucky
Frazer Mining Company, Marion, Kentucky
Hickman, Williams & Company, New York, New York

Minerva Company, Eldorado, Illinois
Ozark-Mahoning Company, Tulsa, Oklahoma

SALTS OF HYDROFLUORIC ACID

Fluorspar is the basic raw material in the manufacture of all fluoride compounds except the silicofluorides, which are generally obtained from the phosphate-rock industry (see page 74). After the acidulation of fluorspar (see page 63) the hydrofluoric acid (HF) formed either can be used directly as a fluoridating chemical or can be employed as the starting compound for many fluoride-containing salts. For instance, by choosing the appropriate reacting compound, almost any fluoride salt can be formed with this acid:

$$HF + NaOH \rightarrow NaF + H_2O$$
$$2HF + Na_2CO_3 \rightarrow 2NaF + H_2CO_3$$
$$2HF + K_2CO_3 \rightarrow 2KF + H_2CO_3$$
$$2HF + CaO \rightarrow CaF_2 + H_2O$$
$$HF + BaOH \rightarrow BaF + H_2O$$

and many others.

Hydrofluoric acid, a liquid, can theoretically be fed directly into water with relatively inexpensive proportioning pumps. Because of the extreme corrosiveness of this acid, however, it is very difficult to handle safely and accurately. At the only community which ever used it for water fluoridation (Madison, Wisconsin), 70 per cent hydrofluoric acid was used in order to conserve space in a series of small well-pump houses. The acid was fed indirectly into the water with a solution feeder. Instead of pumping the acid, the feeder forced an inert mineral oil into the top of the acid tank. The oil, riding on top of the acid, displaced a volume of acid equal to the amount of oil pumped and thereby forced the acid into the water being treated. Corrosion difficulties in piping and fittings were reported to have been overcome by using monel metal or hard rubber. Because of its corrosiveness and the danger inherent in its handling, hydrofluoric acid should not be used in water plants unless the most expert supervision is available. For these reasons, hydrofluoric acid will not be considered in this manual as a material generally suitable for water fluoridation.

SODIUM FLUORIDE

Sodium fluoride is the only fluoride compound presently used for the fluoridation of municipal water from the group of compounds formed from hydrofluoric acid. Some home fluoridation installations involving very small supplies are using potassium fluoride when simultaneous chlorination of the water is practiced.

As shown in Table 7.3, sodium fluoride is a relatively expensive source of fluoride, but because of its unique solubility characteristics it is particularly desirable in certain fluoridation installations. It is almost constantly soluble at about 4 per cent by weight. In other words, about 4 lb sodium fluoride form a saturated solution with 100 lb (12 gal) water. This means that no matter what the temperature of the water in a treatment plant may be, a saturated solution can be formed at about 4 per cent strength. Ordinarily the solubility of salts varies tremendously as the temperature of the water changes. The unusual behavior of sodium fluoride in this regard is utilized in some equipment (described in Chap. 8) that produces such a saturated solution automatically and continuously. This device therefore eliminates the necessity of weighing the compound or measuring the amount of water used to make up a solution of known strength.

Sodium fluoride was used at all of the earlier fluoridation installations (Grand Rapids, Newburgh, and Brantford). This compound was chosen because studies on its toxicity and physiological effects had been completed; it was commercially available in large quantities; it was convenient to use and had a solubility which was relatively high and constant over a wide temperature range. Lately, additional studies have shown that the availability of the fluoride ion for caries inhibition is comparable in sodium silicofluoride, hydrofluosilicic acid, and sodium monofluorophosphate.[2]

Sodium fluoride is a white, odorless, free-flowing material available either as a powder or in the form of minute crystals. Its formula weight is 42.00, specific gravity 2.79, and its solubility practically constant at 4.0 g per 100 ml water at temperatures generally encountered in water-treatment practice. The pH (hydrogen-ion concentration) of solution varies with the source and type of impurities. As purchased

---

[2] I. Zipkin and F. J. McClure, Complex Fluorides: Caries Reduction and Fluorine Retention in the Bones and Teeth of White Rats, *Public Health Repts.* (U.S.), **66**(47): 1523–1532 (1951).

today, sodium fluoride produces solutions with a pH close to neutrality. It is available in purities ranging from 90 to 98 per cent, the impurities consisting of water, free acid (or alkali), sodium silicofluoride, sulfites, and iron. Tentative specifications of the American Water Works Association suggest that:

> Insoluble matter should not exceed 0.50 per cent
> Moisture should not exceed 0.50 per cent
> Free acid should not exceed 0.20 per cent
> Free alkali should not exceed 0.25 per cent
> Fluosilicate should not exceed 0.35 per cent
> Sulfite should not exceed 0.01 per cent
> Heavy metals should not exceed 0.003 per cent
> Iron should not exceed 0.07 per cent
> Other should not exceed 3.12 per cent

Depending on the source and other factors, the density of powdered sodium fluoride varies from a light grade (less than 65 lb per cu ft) to heavy (90 lb per cu ft). The sieve analysis of a material weighing 65 lb per cu ft is 99 per cent through 200 mesh and 97 per cent through 325 mesh. A crystalline grade which contains practically no particles capable of forming dust and which is ideally suited for the sodium fluoride saturator tank described before has a mesh size of approximately 100 per cent through 20 mesh and only 2 per cent through 60 mesh. This material can also be used in dry feeders (volumetric or gravimetric), but generally the powdered grades (99 per cent through 200 mesh) are preferred when rapid and continuous dissolving is required. Any grade of sodium fluoride can be used to make solutions of less than saturated strength (2 per cent or so). In these instances, the preparation of the solutions requires careful weighing of the sodium fluoride and measurement of the water together with prolonged stirring to assure that all of the compound is dissolved. Some state regulations have required that sodium fluoride be tinted a blue color in order to distinguish it from other water-treatment chemicals. This raised the cost of the material approximately 1 cent a pound. The tinted material is manufactured primarily to comply with Federal requirements when it is used as an ingredient of insecticides.

Sodium fluoride is also used as a frosting agent in glass manufacture, as an insecticide, rodenticide, and fungicide, as a preservative of wood and of glue and starch adhesives, and as an ingredient of vitreous enamel, welding and brazing fluxes, and coated papers.

Paper bags (100 lb net), fiber drums (125 lb net), and drums or barrels (375 to 425 lb net) are used for shipping sodium fluoride. The actual dimensions of these containers vary considerably, depending on the manufacturer of the containers. Bulk shipments are now permitted under the Consolidated Freight Classification issued January 26, 1953. Containers are required to carry a general precautionary warning label to avoid hazardous handling.

The following companies manufacture sodium fluoride:

Aluminum Company of America, Pittsburgh, Pennsylvania
American Agricultural Chemical Company, New York, New York
Blockson Chemical Company, Joliet, Illinois
General Chemical Company, New York, New York

POTASSIUM FLUORIDE

At the present time there are no municipal water plants using potassium fluoride. It is used only at a few experimental installations where individual home water systems are being fluoridated. This compound was chosen because such water systems were being chlorinated at the same time and with the same equipment used for adding fluorides. The solution containing the fluorides also contained the chlorine in the proportion desired to give the optimum fluoride concentrations and chlorine residuals. The characteristics of this compound are shown in Table 7.3. Inasmuch as only the purest grade (reagent) of potassium fluoride has been used, it is considerably more expensive than the others. Other compounds containing fluorides could also be used for this purpose (i.e., lithium fluoride, magnesium fluoride).

SILICOFLUORIDES

Source and Preparation

Practically all the commercially available silicofluorides are obtained as a by-product of the purification of phosphate rock. The primary products are superphosphates, phosphoric acid, elemental phosphorus, and triple superphosphate. By far the largest proportion of these products is used in the preparation of chemical fertilizers.

Of the many sources of phosphates (bones, guano, slag) the most important are the phosphate-containing rocks (phosphorites) found in many parts of the world. They generally occur as sedimentary deposits, which are usually of marine origin. Such deposits always contain fluoride (ranging from 2.9 to 6.9 per cent by weight) and an oxide of phosphorus in ratios which approach the composition of apatites [$Ca_{10}(PO_4,CO_3)_6(F,Cl,OH)_2$]. The presence of fluorine as a part of the phosphate molecule is probably responsible for the low solubility of the rocks and accounts for the accumulation and preservation of the enormous quantities found in some areas.

The most important deposits of this mineral are in the United States and include the extensive reserves in Florida, Tennessee, and some Western states, notably Montana, Idaho, Utah, and Wyoming. Other large areas are in North Africa, including Algeria, Tunis, Morocco, Egypt, in the U.S.S.R., and on various islands in the Indian and Pacific Oceans. Smaller deposits are found in Belgium, France, England, Australia, Japan, New Zealand, South America, and South Africa. A somewhat typical analysis of a phosphate rock (Florida) before beneficiation would be:

Calcium oxide (CaO)	46.5%
Carbon dioxide	3.5
Silica (SiO_2)	9.5
Phosphoric acid (P_2O_5)	34.0
Iron, aluminum, and other oxides	3.0
Fluorine	3.5
	100.0%

The known domestic phosphate reserves as of 1950 amounted to more than 13 billion tons, which contained 420 million tons of fluorine (averaging over 3 per cent fluorine). The fluorine from this source alone would therefore be sufficient for supplying the total requirement for all the public water supplies in the United States for the next 27,000 years or so. The production of phosphates in 1961 consumed 17.5 million tons of marketable (salable products from rock-purification plants) phosphate rock containing about 525,000 tons of fluorine. Only about 3.0 per cent (15,000 tons) of this amount of fluorine would be required to fluoridate all the public water supplies in the United States.[3] Some of the products of the phosphate-rock

[3] W. L. Hill and K. D. Jacob, Phosphate Rock as an Economic Source of Fluorine, *Mining Eng.*, 6(10): 994 (1954).

industry and the quantities involved for the year 1961 are shown in the accompanying table.

Uses	Millions of long tons	Per cent of total
Agricultural:		
Superphosphate...............................	4.4	25
Triple superphosphate..................	4.6	26
Stock feed and other.....................	0.9	6
Total..	9.9	57
Industrial:		
Elemental phosphorus, phosphoric acid....	3.7	21
Exports and other uses..................	3.8	22
Total..	7.5	43

Of this total of 17 million tons of phosphate products produced, approximately 75 per cent originated in Florida.[4] Most of the fluorine from this source is available from the processes involving the production of superphosphates.

From the agricultural point of view, the production of superphosphates involves the conversion of insoluble tricalcium phosphate, which is usually present as a fluorapatite in the phosphate-rock deposit, into a soluble phosphate salt and calcium sulfate. This is done by dissolving the phosphate rock in sulfuric acid after the gross impurities are removed. The process is generally carried out in four steps:

1. Grinding the phosphate rock preparatory for acid treatment
2. Mixing and reacting the finely ground rock with sulfuric acid
3. Curing and drying the mixture of acid and rock
4. Milling and bagging the finished material

The evolution of fluorine is restricted predominantly to the mixing and reacting operations in a closed chamber or "den." The denning operation provides an intimate commingling of the acid and rock with various means for conserving the heat of the reaction and controlling the gases evolved. Most modern dens are continuously operated, the acid and rock being fed automatically to the den, from which the mixture is withdrawn by conveyors to the curing piles.

When acid is added to finely ground tricalcium phosphate, calcium

[4] R. W. Lewis and G. E. Tucker, Mineral Market Report No. 3347, December, 1961, Division of Minerals, Bureau of Mines, Department of the Interior.

sulfate and monocalcium phosphate are formed according to the following reaction:

$$Ca_3(PO_4)_2 + 2H_2SO_4 + H_2O \rightarrow 2CaSO_4 + CaH_4(PO_4)_2 \cdot H_2O$$

Impurities contained in the tricalcium phosphate also consume sulfuric acid. The fluorine, in the form of calcium fluoride, reacts somewhat as follows:

$$CaF_2 + H_2SO_4 \rightarrow CaSO_4 + 2HF\uparrow$$

The hydrofluoric acid reacts with the silica impurities and assists the sulfuric acid in dissolving the rock:

$$4HF + SiO_2 \rightarrow SiF_4\uparrow + 2H_2O$$

The gas evolved, silicon tetrafluoride, is reacted with water from which is obtained hydrofluosilicic acid and the insoluble silica:

$$3SiF_4 + 2H_2O \rightarrow SiO_2 + 2H_2SiF_6$$

A large portion of the silicon tetrafluoride formerly escaped without combining with water and was discharged from the plant with the stack gases. This created a considerable nuisance in the neighborhood of the plant and a hazard to the employees. It is customary now to absorb the gas in water (to form hydrofluosilicic acid) or through beds of limestone to form calcium fluoride. At present only a minor portion of the fluorides are converted to hydrofluosilicic acid—only about 29 per cent of the fluorine in the Florida rock is evolved, the rest remaining in the superphosphate. Of this 29 per cent, only about one-sixth is recovered for fluosilicic acid production, the rest being diluted and discharged as waste.[5] This very low ratio of recovery is predominantly due to the lack of economic incentive for producing silicofluorides and to the ease with which the waste can be disposed of at the present time. Because of the increased awareness of the deleterious effects of stream and air pollution, it may well be that a much larger proportion may be recovered in the future.

HYDROFLUOSILICIC ACID

Hydrofluosilicic acid (fluosilicic acid, hexafluosilicic acid, silicofluoric acid) is a 20 to 35 per cent aqueous solution of H_2SiF_6 with a

[5] William H. Haggaman, "Phosphoric Acid, Phosphates, and Phosphatic Fertilizers," Reinhold Publishing Corporation, New York, 1952.

formula weight of 144.08. It is not known in the anhydrous form (containing no water). It is a colorless, transparent, fuming, corrosive liquid having a pungent odor and an irritating action on the skin. A 22 per cent solution boils at about 221°F and freezes at about 4°F. A 1.0 per cent solution has a pH of 1.2. Densities at 17.5°C for various concentrations are:

H_2SiF_6, per cent	6	14	22	25	30	34
Densities	1.0491	1.1190	1.1941	1.2235	1.2742	1.3126

Upon vaporizing, the acid decomposes to hydrofluoric acid (2HF) and silicon tetrafluoride. This equilibrium exists at the surface of strong solutions of this acid, and the small concentration of hydrofluoric acid may slowly attack the glass container above the solution level. Rubber-lined containers are therefore generally used for shipping and storage.

This acid is made from gases which are generated during the acidulation of phosphate rock and which contain low concentrations of silicon tetrafluoride and water vapor. These are drawn by fans through a tower, which is equipped with trays. Here the gases are absorbed in recirculated dilute hydrofluosilicic acid at about 160°F. When the recirculating acid reaches the desired strength, the system is discharged to filters, where some of the free silica is removed. The ducts, fans, tower, and filters are lined with neoprene. The acid pump is monel-fitted. Tanks for storing the cold acid are rubber-lined. The maximum concentration attained by this method is 30 per cent. Higher concentrations of this acid are made by dissolving silica (sand) in hydrofluoric acid. The higher cost of this more concentrated acid is, however, disproportionate from the standpoint of fluoride content when compared to the cost of the weaker product. Generally the acid made from silicon tetrafluoride is of satisfactory strength for water fluoridation.

While this acid is only slightly volatile, care should be observed in avoiding the vapor. Inhalation of silicon tetrafluoride can cause irritation of the respiratory tract. Equipment exposed to the fumes becomes corroded. For these reasons, storage tanks should be covered and, if placed indoors, adequate ventilation should be provided. Volatility of the acid can be reduced by maintaining it at as low a temperature as possible.

Fluoride Compounds (Characteristics, Sources, Costs)

Compared with hydrofluoric or sulfuric acids, this acid is relatively safe to handle. Nevertheless, as soon as possible after exposure, it should be washed off the hands or other parts of the body with liberal amounts of water. Protective clothing (as described in Chap. 12) and goggles should be worn while handling it.

Because of its corrosive characteristics, care should be taken in selecting materials which will withstand deterioration. The materials of construction for tanks, pipe, pumps, valves, and other appurtenances are arranged according to their relative suitability in Table 7.1.

TABLE 7.1. CORROSION RESISTANCE OF VARIOUS MATERIALS

Excellent	Fair	Unsuitable
Hastelloy C	Copper	Wood
Structural carbon	Brass	Mild steel
Durimet 20	Monel	Stainless steel
Neoprene	Hard rubber	Lead
Natural rubber	Wood (pitch-lined)	
Saran	Glass	
Teflon	Ceramics	
Polyethylene		
Kel-F		

The American Water Works Association Standard Specification for Fluosilicic Acid (AWWA Specification B703) limits the total impurities of lead, arsenic, and antimony to 0.04 per cent. No limit is placed on silica, although this element is sometimes present in appreciable quantities. If an excess of free silica occurs, a visible, insoluble precipitate forms when the dilution ratio is less than 20:1. In other words, there is no visible precipitate when more than 20 parts of water are added to 1 part of acid. Dilution of the acid with water is sometimes necessary at some small water plants because of the limitations in accurate delivery of chemical solution feeders when set at very low rates. This precipitate clogs pumps, valves, pipelines, and other fluoride feeding equipment. It has been found[6] that the addition of small amounts of hydrofluoric acid to the fluosilicic acid will prevent the formation of this precipitate, which is finely divided silica. The amount added is in the range of 1 gal hydrofluoric acid (48 per cent) to 50 gal fluosilicic acid (30 per cent). The optimum ratio de-

[6] E. Bellack and F. J. Maier, Dilution of Fluosilicic Acid, *J. Am. Water Works Assoc.*, **48**(2): 199 (1956).

pends on the quantity of colloidal silica which must be dissolved; this is best determined by the manufacturer. In the interest of safety at the water plant and for maximum effectiveness, manufacturers should stabilize the acid at their plants by adding the optimum quantities of hydrofluoric acid.

It is generally not advisable, however, to choose fluosilicic acid if it must be diluted prior to feeding it. If this step is necessary, it indicates that the water plant involved requires relatively small amounts of chemicals with a resulting higher cost of available fluoride from fluosilicic acid compared with other sources. In addition, the dilution requires accurate measurement of both the acid and the dilution water. The necessity for these measurements contributes to possible errors in the strength of the resulting mixture. By and large, when small plants are involved, it is more economical and easier to use sodium fluoride solutions obtained from a saturator.

Table 7.3 indicates that of all the compounds now being used for public water supplies, this acid is the most expensive source of fluoride. Table 7.2, however, indicates that except for sodium silicofluoride, this acid serves more people for fluoridation than does any other compound. This is primarily because three of the largest cities in the United States (Chicago, Baltimore, and Philadelphia) are using it. They are able to obtain it at a very favorable price from local producers. In addition, because the acid is invariably fed and proportioned into the water with relatively inexpensive solution feeders, the capital cost of hydrofluosilicic acid installations is generally considerably less than any other.

In addition to fluoridation, this acid is used for making silicofluorides and fluoride salts; as an agent for increasing the hardness of china, porcelain, and pottery; as an ingredient of disinfectants and paint; as a wood preservative; as a bath in the electrolytic refining of lead and the electroplating of chromium; for surface cleaning aluminum and treating glass surfaces to reduce reflection.

Hydrofluosilicic acid is priced on the basis of 30 per cent H_2SiF_6 solution even though the weaker strength may be actually (and generally) shipped. It is classed as a corrosive liquid by the Interstate Commerce Commission and shipped under a white acid label, manifested as "hydrofluosilicic acid." Samples cannot be shipped by mail, and when shipped by express, they must be packed in such a manner as to allow absorption of the acid in the dunnage if the container should leak. It is shipped in 160- or 500-lb wooden barrels or in 125-lb rubber-lined drums (ICC Specification 43A). Deposit is required on drums (approximately $75 on 500-lb drums). Some manufacturers also have available rubber-lined tank cars and trucks of capacities up to about 90,000 lb net. Fluosilicic acid is manufactured by:

American Agricultural Chemical Company, New York, New York
Baugh & Sons Company, Philadelphia, Pennsylvania
Davison Chemical Company, Baltimore, Maryland
E. I. Du Pont de Nemours & Company, Inc., Wilmington, Delaware
Harshaw Chemical Company, Cleveland, Ohio
Tennessee Corporation, Atlanta, Georgia

SODIUM SILICOFLUORIDE

Hydrofluosilicic acid is the basic raw material used in the manufacture of silicofluoride salts. There are many of these made, but so far only sodium silicofluoride, magnesium silicofluoride, and ammonium silicofluoride have been used for water fluoridation.

Of all the compounds used for this purpose, sodium silicofluoride is by far the most popular. As shown in Table 7.2, the water supplies

TABLE 7.2. ANNUAL CUMULATIVE POPULATION SERVED FLUORIDATED WATER, BY CHEMICAL USED, 1945–1960

Year	Total population	Chemical					
		Sodium fluoride	Sodium silicofluoride	Fluosilicic acid	Ammonium fluosilicate	Calcium fluoride	Other, adjusted natural fluoride, and not specified
1945	231,920	231,920					
1946	332,467	315,747	16,720				
1947	458,748	428,028	30,720				
1948	581,683	430,963	30,720	120,000			
1949	1,062,779	840,225	96,274	120,000	......	.....	6,280
1950	1,578,578	990,764	376,211	200,774	......	.....	10,829
1951	5,079,321	1,401,942	3,354,055	305,484	7,011	.....	10,829
1952	13,875,005	2,485,054	9,461,388	1,901,723	16,011	.....	10,829
1953	17,666,339	2,991,190	12,505,418	1,957,697	17,066	.....	194,968
1954	22,336,884	2,740,248	15,138,850	4,126,862	59,621	.....	271,303
1955	26,278,820	3,608,006	17,252,295	5,001,934	60,636	2,700	353,249
1956	33,905,474	3,922,181	19,656,155	9,792,003	61,611	3,160	470,364
1957	36,215,208	4,302,768	21,265,391	9,882,735	62,626	3,620	698,068
1958	38,461,589	4,456,953	22,919,696	10,083,504	63,600	6,749	931,087
1959	39,628,377	4,626,851	23,851,411	10,146,814	64,696	7,190	931,415
1960	41,169,412	5,186,351	24,588,937	10,285,530	65,710	7,224	1,035,660
1961	42,201,115	5,360,543	25,092,573	10,626,715	66,367	7,296	1,047,621

TABLE 7.3. CHARACTERISTICS OF FLUORIDE COMPOUNDS

Item	Calcium fluoride (CaF_2)	Sodium silicofluoride (Na_2SiF_6)	Sodium fluoride (NaF)	Hydrofluosilicic acid (H_2SiF_6)	Magnesium silicofluoride ($MgSiF_6 \cdot 6H_2O$)	Ammonium silicofluoride ($(NH_4)_2SiF_6$)	Potassium fluoride ($KF \cdot 2H_2O$)
Form	Powder	Powder	Powder	Liquid	Crystal	Crystal	Crystal
Molecular weight	78.08	188.05	42.00	144.08	274.48	178.14	94.13
Commercial purity, %	85–98	98.5	90–98	22–30	98	98	98
Fluoride ion, % (100% pure material)	48.8	60.7	45.25	79.2	41.5	63.9	20.2
Weight (lb per cu ft)	101	55–72	65–90	10.5 lb/gal	73	81	58
Storage space, cu ft per 1,000 lb F ion	21–24	23–30	22–34	54–73	34	21	87
Lb required per mg for 1.0 ppm F at indicated purity	17.6 (97%)	14.0	18.8 (98%)	35.2 (30%)	20.5	13.3	42.1
Solubility (g per 100 g H_2O) at 25°C (77°F)	0.0016	0.762	4.05	Infinite	64.8 (17.5°C)	18.5 (17.5°C)	100
Gal. saturated solution required per mg water at 25°C and 1.0 ppm F	52,500	219	55.6 (98%)	3.35	3.5	8.0	8.0
pH of saturated solution	6.7	3.5	7.6	1.2 (1% sol)	1.0	3.5	7.0
Cost:							
Cents per lb	2.2	7.5	14.4	10.5 (30%)	13.0	11.75	37.5
Cents per lb available F	4.8	12.6	32.5	44.2	32.0	18.8	189.0
Per mg for 1.0 ppm	$ 0.40	$ 1.05	$ 2.72	3.68	$ 2.66	$ 1.56	$ 15.77
Per year at 1.0 mgd	$146.00	$383.00	$990.00	$1,340.00	$973.00	$570.00	$5,760.00

Fluoride Compounds (Characteristics, Sources, Costs)

treated with sodium silicofluoride serve more people than do those treated with all the other fluoride compounds combined. The primary reason for this is its low cost. As shown in Table 7.3, sodium silicofluoride is the cheapest source of fluoride ion except for fluorspar. While its solubility is difficult, this drawback can be overcome by selecting the correct dissolving apparatus. There are some installations feeding slurries (solution of a salt of such excessive strength that not all the salt can dissolve) of sodium silicofluoride in order to

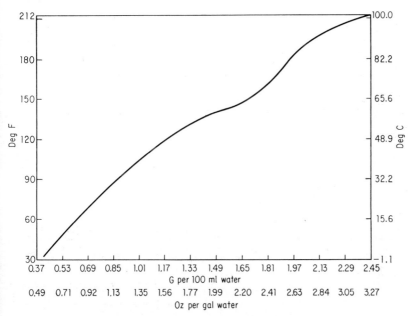

FIG. 7.5. Solubility of sodium silicofluoride. (USPHS.)

save money on dissolving equipment. This procedure is not ordinarily recommended, since undissolved particles of the compound may be fed into the distribution system or may sink to the bottom of a storage tank and be lost.

Sodium silicofluoride (sodium fluosilicate) is a white, free-flowing, odorless, nonhygroscopic crystalline powder. Its molecular weight is 188.05, density 2.679, density of saturated solution 1.0054, and pH of a saturated solution 3.5 to 4.0. The hydrolysis permits direct analytical determination by titration with a standard alkaline solution. The solubility at various temperatures is shown in Fig. 7.5. Its limited solubility makes it valuable as a laundry sour, since there is no danger of the acidity becoming great enough to damage fabrics.

The sparingly soluble fluosilicates of barium, potassium, and sodium are precipitated when a soluble salt of one of these metals is added to hydrofluosilicic acid:

$$H_2SiF_6 + 2NaCl \rightarrow Na_2SiF_6\downarrow + 2HCl$$

In practice, brine is added to 25 per cent hydrofluosilicic acid in rubber-lined tanks equipped with agitators, centrifuging the slurry and drying the solid sodium silicofluoride.

Two grades of fineness are commercially available—regular (72 lb per cu ft) and fluffy (55 lb per cu ft). A typical sieve analysis of the regular grade is 100 per cent through 40 mesh and 95 per cent through 100 mesh; for the fluffy grade, 99.5 per cent through 200 mesh and 97.5 per cent through 325 mesh. That material intended for insecticide use is tinted nile blue. A typical chemical analysis of a satisfactory grade is:

Na_2SiF_6 .... 98%+
Water 0.5%
SiO_2 .... 0.1%
Chlorides 0.6%
Heavy metals .... Less than 0.05%

The present American Water Works Association specifications do not limit the heavy-metal content. Both insoluble matter and moisture are limited to 0.5 per cent.

Sodium silicofluoride is used in rodenticides, insecticides, fungicides, and bactericides, as a coagulant for latex, as a laundry sour, in the manufacture of enamels and glass, in the pretreatment of hides and skins, and in the extraction of beryllium. Heretofore its only use in the water-supply industry has been as a bactericide in sulfur-base pipe-joining compounds.

Sodium silicofluoride is rarely proportioned into water as a solution because of the difficulty in dissolving it. Instead it is almost invariably introduced by means of dry feeders into the solution box or dissolver, which is a part of all dry feeders. Some grades of sodium silicofluoride have not been found to be capable of accurate feeding, since the material arched in the storage hopper or the feeders, or ran through the feeding mechanism uncontrolled (similar to the action of a liquid). This is believed to have been caused by either the size distribution of the particles or an excess of water in the material. Based on reports from water plants on the "feedability" of the various grades of material, present belief is that the water content should not exceed about 0.5 per cent and the size should be well within the limits of 100 and 325 mesh. A "feedability" index has been suggested (unpublished data, Division of Dental Public Health, U.S. Public Health Service) in which both moisture and size are considered. If this index exceeds 80, the material is found to be satisfactory. The formula is $F = 100 - (A + B + 10C)$ in which F is the "feedability" index; A is the percentage retained on 100-mesh sieve; B is the percentage passing through 325-mesh sieve; and C is the percentage of moisture. For example, a sample of sodium silicofluoride reported to be satisfactory tested 3 per cent on 100 mesh, 9 per cent passing 325 mesh, and 0.05 per cent moisture. Therefore:

$$F = 100 - (3 + 9 + 0.5) = 87.5$$

Since solutions of sodium silicofluoride are corrosive, materials used for pumps, valves, pipes, etc., should be chosen accordingly. The materials listed for hydrofluosilicic acid on page 79 are also satisfactory for solutions of sodium silicofluoride.

The effectiveness of sodium silicofluoride as a fluoridation agent has been investigated[7] and shown to be similar to that of sodium fluoride. Its use in communities throughout the country has also shown it to be a very practical and effective material for this purpose.

Paper bags (100 lb net), paper-lined barrels (425 lb net), and wooden fiber drums (125 lb net) are used for shipping sodium silicofluoride. Bulk shipments are now permitted by the Consolidated Freight Regulations. Dimensions of the containers are: paper bags: 16½ × 13 × 5 in.; drums: 13 × 22 in.; barrels: 23 × 30 in. Containers carry a general precautionary warning label to avoid hazardous handling.

Sodium silicofluoride is manufactured by and obtainable from:

American Agricultural Chemical Company, New York, New York
Baugh Chemical Company, Baltimore, Maryland
Blockson Chemical Company, Joliet, Illinois
Davison Chemical Company, Baltimore, Maryland
E. I. Du Pont de Nemours & Company, Inc., Wilmington, Delaware

[7] F. J. McClure, Availability of Fluorine in Sodium Fluoride vs. Sodium Silicofluoride, *Public Health Repts.* (*U.S.*), **65**(37): 1175 (1950).

Henry Sundheimer Company, New York, New York
Tennessee Corporation, Atlanta, Georgia
U.S. Phosphoric Products, Inc., Tampa, Florida

AMMONIUM SILICOFLUORIDE

Ammonium silicofluoride is produced by neutralizing fluosilicic acid with either aqueous ammonia or ammonia in the gaseous form. The product of this reaction is centrifuged to remove most of the water and is then oven-dried. It is sold as a white, odorless, free-flowing crystalline material containing few particles capable of producing a dust. A typical mesh size is 100 per cent through 20 mesh and only 10 per cent through 100 mesh.

Ammonium silicofluoride has a formula weight of 178.14, and the 98 per cent pure material contains 62.7 per cent fluoride, 15.4 per cent silica, and 18.75 per cent ammonia. Its solubility is high—up to 55.5 per cent at 100°C, but it falls to 13 per cent at 0°C. Its apparent specific gravity is 2.1 and the pH of a saturated solution is 3.5. As shown in Table 7.3, in cost it compares favorably with sodium fluoride.

The use of this material is particularly desirable wherever ammonia is used to form chloramines with the chlorine added to the water for disinfection purposes. Chloramines are preferred to chlorine as a disinfectant of water when the pH is low enough and the period of contact with the water sufficiently prolonged (i.e., where a reservoir or long transmission main is used between the point of treatment and use). There is only a small number of supplies where these conditions exist. The use of ammonium silicofluoride under these conditions provides part or all of the ammonium required. Theoretically, to form chloramines, the chlorine required is four times the amount of ammonia fed. Inasmuch as 13.3 lb ammonium silicofluoride is used to produce 1.0 ppm fluoride in 1 million gal water, then 2.5 lb ammonia would be supplied. There would then have to be fed at least 10 lb chlorine per million gal water to form chloramines with this quantity of ammonia. A free residual chlorine would then be produced by any amount of chlorine in excess of 10 lb per million gal.

Ammonium silicofluoride is also used for mothproofing textiles, in the sand-mold casting of magnesium, and as a laundry sour. As shown in Table 7.2, ammonium silicofluoride is used by about 60,000 people living in three communities.

This material is available in 100-lb and 400-lb drums and, of course, in 1-lb packages if desired. It can also be obtained in carload or truckload lots. It is manufactured by Daniel H. Jones Laboratories, Camden, New Jersey, and Davison Chemical Company, Baltimore, Maryland.

MAGNESIUM SILICOFLUORIDE

When hydrofluosilicic acid is neutralized with either magnesium carbonate or magnesium hydroxide, the resulting salt is magnesium silicofluoride. At present this compound is used only for demonstration home fluoridation installations, although it could be readily used for municipal water supplies. As shown in Table 7.3, it is more soluble (except for potassium fluoride) than any other fluoride compound used for this purpose. Its cost is well within the range of sodium fluoride, and where dry feeders are used, it could perhaps be used in place of sodium fluoride with a resulting saving in chemical costs. It is available in 50-lb cartons from the American Agricultural Chemical Company, New York, New York.

CRYOLITE

Of the three more commonly found minerals containing the largest percentage of fluoride (fluorspar, apatite, and cryolite), only cryolite has no significance in the fluoridation of water. Cryolite, Na_3AlF_6, has the highest fluorine content of any known commercial mineral, 54 per cent by weight. It is mined only at Ivigtut, Greenland, but deposits have been reported at Pikes Peak and in the Ural Mountains in Russia. Because of its cost, it is not used as a raw material in the fluorine chemical industry. It is employed primarily in the reduction of aluminum, although much of the cryolite used for this purpose is synthetic, being manufactured from fluorspar.

SELECTION OF THE BEST FLUORIDE COMPOUND

The foregoing review of the chemical and physical characteristics of various fluoride compounds is included to assist prospective users in selecting the one best material for a particular installation. The cost, availability, and quality will vary from place to place so that the final choice must be based on a more or less detailed study of the advantages and disadvantages of each compound locally available. The various characteristics that are ordinarily considered are discussed in the following sections.

Chemical Costs

Based on the cost of the available fluoride ion, fluorspar is by far the least expensive source. It is believed that whenever coagulants are used for water treatment, means might be found for utilizing this material for fluoridation. When fluorspar cannot be used, then sodium silicofluoride should next be considered. After this, in order of cost, should be placed ammonium silicofluoride, magnesium silico fluoride, sodium fluoride, and hydrofluosilicic acid.

Solubility

If 1 gal hydrofluosilicic acid is required to fluoridate a certain quantity of water, then 17 gal of a saturated solution of sodium fluoride, or 65 gal of a saturated solution of sodium silicofluoride, will be required for the same amount of water. This inherent advantage in hydrofluosilicic acid is reflected in the lower cost of solution feeders and solution tanks, which can be used with this acid. The advantage of the acid's high solubility, however, is lost in the larger installations, where the other less expensive compounds are proportioned directly with dry feeders.

Chemical Storage Space

For the same quantity of fluoride ion, sodium silicofluoride requires the least storage space; hydrofluosilicic acid the most. However, the acid is stored in tanks, which might be placed outdoors or underground. Practically speaking, consideration of storage requirements as a factor in choosing a particular compound depends primarily on a comparison between costs for providing additional storage space and the price of the compounds.

Feeder Space Limitations

Hydrofluosilicic acid has been chosen in some communities where available space in small treatment plants is limited. Inasmuch as the acid can be stored in tanks at some distance from the feeder and no solution make-up equipment is required, the use of the acid is advantageous when room for only the solution feeder is available.

Corrosiveness

Although the relative corrosiveness of fluoride compounds differs, equipment manufacturers provide about the same corrosion-resisting materials for solution feeders, tanks, and appurtenances regardless of the compound used. About the only exception is the action of hydrofluosilicic acid on ceramic crocks or glass-lined tanks (as a result of vaporization and the formation of small amounts of hydrofluoric acid on the liquid surface). Hydrofluosilicic acid will seldom be used in such containers, however, because it can be drawn directly from the rubber-lined drums in which it is shipped.

Incrustation in Feeding Equipment

Incrustants form occasionally in solution tanks, feeders, and feeder lines. These are essentially calcium and magnesium fluoride or silicofluoride derived from strong fluoride solutions and the hardness constituents of the water. No incrustants are formed when hydrofluosilicic acid is used because it is not ordinarily diluted with water. Incrustants formed when sodium silicofluoride (calcium or magnesium silicofluoride) is used are slightly more soluble than calcium and magnesium fluoride. The problem can be minimized by softening the solution water or applying small amounts of hexametaphosphates.

The use of softeners to reduce loss of fluorides caused by fluoride precipitation is not ordinarily justified. The saving realized over the cost of softening amounts to less than $2 per year in a city of 10,000 population using water of 200 ppm hardness.

There is very little additional labor involved in removing the precipitated fluorides from solution or saturator tanks, inasmuch as sodium fluoride contains insoluble material which also must be removed periodically.

Bulk Chemical Handling

Relatively little manual labor is required in treatment plants using hydrofluosilicic acid, since this material can be readily pumped. In the larger plants, where carload shipments in bulk can be used, the dry powdered or crystalline materials can be easily and economically handled with mechanical or pneumatic conveyors. Considerable effort is required in the medium-sized plants receiving their chemicals in barrels or bags. These must be moved about the plant and emptied into the feeder hoppers or solution tanks, generally by hand. In general, hydrofluosilicic acid requires the least handling in a plant of any size, while the effort required in handling the dry material depends on the size of the plant.

Hazards to Operators

Dust rising from the handling of powdered fluorides can be controlled by careful handling and the use of protective devices. Practically no hazards are involved in using the acid, since it is ordinarily proportioned directly from the shipping container. At some of the smaller plants, where the acid requires dilution prior to feeding, considerably more care in handling is required. No particular advantage is gained by this procedure, however, because about as much effort is required to dilute the acid as to make up solutions of the less expensive dry materials.

CHAPTER 8　*Feeder Types and Capacities*

A fluoridation installation must be considered as a complete process and, as such, its various components cannot be chosen without considering their effect on the other parts. For instance, each of the characteristics of fluoride compounds described in Chap. 7 will have, to some extent, a direct bearing on the selection of the best feeder to use, its location, the point of fluoride application, and the type of auxiliary equipment.

The best installation is one that incorporates the most desirable combinations of these factors:

1. Simple, accurate feeding equipment
2. Minimum chemical handling
3. Consistent with the above two factors, the lowest overall cost based on amortization of equipment and cost of chemical
4. Ease in collecting reliable records
5. Minimum maintenance of feeder, piping, and injector equipment

Devices for feeding fluorides accurately have generally been adapted from those machines originally designed for feeding a variety of liquid or solid chemicals in water-treatment and industrial plants.

Fluorides are proportioned into a water supply either as liquids or solids. Chemical feeders can therefore be broadly divided into two types: (1) solution feeders, which are essentially small pumps, used to feed a carefully measured quantity of accurately prepared fluoride solution (or hydrofluosilicic acid) during a specified time and (2) dry feeders, which deliver a predetermined quantity of the solid material during a given time interval. Dry feeders are further subdivided by types, depending on the method of controlling the rate of delivery. Volumetric dry feeders deliver a measured volume of dry chemical within a given time interval; gravimetric feeders (loss-in-weight) deliver a measured weight of chemical within a given period.

The choice of a feeder depends on the fluoride compound used and the amount to be fed. The rate of feed will depend on the desired fluoride content of the treated water, the amount of water to be treated passing a given point, and the fluoride content of the untreated water. In general, solution feeders are used for the smaller supplies and dry feeders for the larger ones. There is, of course, a wide area within which either type would be equally successful. The capacities of the various feeders for introducing 1.0 ppm fluoride into water at different rates of flow are shown in Table 8.1.

TABLE 8.1. USUAL RANGE OF FLUORIDE FEEDERS

Type of feeder	Chemical used	Feed-rate range		Gpm treated with 1.0 ppm F	
		Minimum	Maximum	Minimum	Maximum
Gravimetric dry feeder	Na_2SiF_6	0.5 lb per hr	5,000 lb per hr	600	6,000,000
	NaF	0.5 lb per hr	5,000 lb per hr	440	4,400,000
Volumetric dry feeder	Na_2SiF_6	1.0 oz per hr	5,000 lb per hr	75	6,000,000
	NaF	1.0 oz per hr	5,000 lb per hr	55	4,400,000
Piston or centrifugal pump	Solution of 22% H_2SiF_6	0.375 ml per min	Unlimited	18	Unlimited
	Solution of 4% NaF	0.375 ml per min	Unlimited	2	Unlimited
Diaphragm pump	Solution of 22% H_2SiF_6	2.0 ml per min	1 gpm	110	210,000
	Solution of 10% $(NH_4)_2SiF_6$	2.0 ml per min	1 gpm	34	65,000
	Solution of 4% NaF	2.0 ml per min	1 gpm	9.5	18,000

SOLUTION FEEDERS

Solutions of fluorides can be fed into a water supply in the following ways:

1. Saturated solutions of sodium fluoride in constant strengths of

Feeder Types and Capacities

practically 4 per cent can be produced in a saturator tank at almost any temperature of water encountered in the usual water plant. A diagram of this device is shown in Fig. 8.1. The tank is made of stainless steel or fiberglass and is equipped with an inner cone connected to a tube through which the suction line from the feeder passes. The cone is covered with a bed of graded sand and gravel on which the crystalline sodium fluoride rests. Up to about 200 lb sodium fluoride is added to the tank together with the water. A float valve automatically maintains a constant water level over the sodium fluoride bed. The water slowly trickles down through the sodium fluoride and becomes saturated with it before it reaches the sand and gravel. It is then withdrawn from the saturator with a solution feeder. The drawing (Fig. 8.1) also shows two tanks which, as explained in Chap. 9, are sometimes included to soften the water.

Some small amount of insoluble material will always collect on top of the bed, and arrangements must be made to remove these periodically, usually about once a year. This can be most readily done by introducing water under pressure down the tube containing the feeder suction hose. The insoluble particles resting on top of the bed will generally rise with the water introduced through the gravel and be removed through the overflow pipe near the rim of the tank.

2. Unsaturated solutions of sodium silicofluoride, sodium fluoride, magnesium silicofluoride, or ammonium silicofluoride are prepared by weighing amounts of the compounds, measuring quantities of water, and thoroughly mixing them together. Three errors are therefore possible: (a) in weighing the compound, (b) in measuring the quantity of water, and (c) in mixing insufficiently. Any one of these errors will result in a solution different from the strength intended and consequently variations in the fluoride content of the treated water.

3. Solutions of hydrofluosilicic acid are used either as delivered (22 to 30 per cent) or, if necessary, diluted with water to a definite strength. Again, as in method 2 above, attempts to dilute the acid are subject to errors in measuring both the acid and the diluting water. It is much better to use the acid undiluted as it comes from the containers in which it is shipped. If the acid is too concentrated for the solution feeder to handle, then weaker solutions of other, less expensive compounds are generally indicated; for instance, saturated solutions of sodium fluoride. If the acid must be diluted, care should be

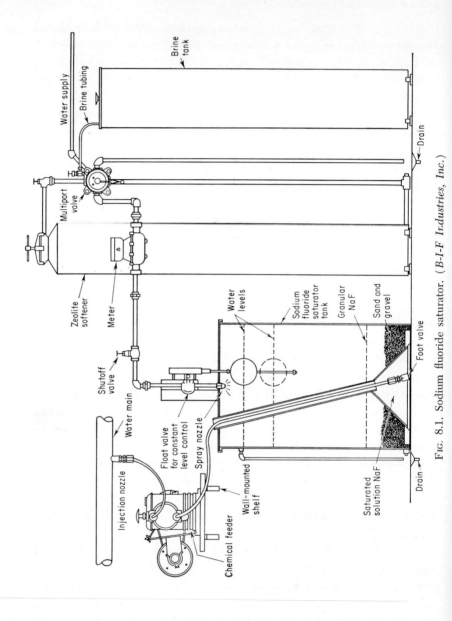

Fig. 8.1. Sodium fluoride saturator. (*B-I-F Industries, Inc.*)

taken to avoid the formation of a precipitate of silica, which will appear despite the quality (hardness) of the water used for dilution. This can be invariably avoided by specifying that the acid should be fortified with hydrofluoric acid to a degree that will dissolve all the uncombined silica.

4. Miscellaneous applications: Solution feeders with special suction and discharge valves can be used for feeding slurries of fluoride compounds. These are mixtures of fluoride compounds (or any other dissolvable or insoluble compound in a solvent, like water) in concentrations greater than saturation. In other words, slurries contain undissolved particles of compound because the water cannot dissolve any more than a certain maximum quantity at a particular temperature. Generally, it is safer and more economical to avoid feeding slurries in fluoridation, because undissolved particles may reach the consumers (if the point of fluoride application is very close to the first customer) and also because if no mixing occurs after the point of application, some of the undissolved particles may settle from the water and remain undissolved.

In two additional systems of fluoridation that have been used, the solutions of fluorides have been fed indirectly with solution feeders. One of these (described in Chap. 7) involved the use of hydrofluoric acid at Madison, Wisconsin. Because of the extreme corrosiveness of this acid, the feeder proportioned mineral oil into a tank containing the hydrofluoric acid. The oil displaced an equal volume of acid so that the acid did not come in contact with the feeder.

Also described in Chap. 7 were two systems for dissolving and feeding solutions of fluorspar. Each of these required, by means of solution feeders, the accurate feeding of alum solutions into a tank containing a bed of fluorspar. The amount of alum solution delivered by these feeders not only controlled the quantity of fluorides dissolved from the fluorspar but also was used to suspend the resulting gypsum and separate it from the powdered fluorspar.

Solution feeders are devices for measuring and delivering a specific volume of liquid during a given time interval. The measuring function is accomplished by the automatic filling of a space in the feeder head with the liquid and then discharging it on the next cycle. This space can be in a pump cylinder, be formed behind a diaphragm, be in a series of buckets attached to a rotating wheel, or be the difference in level in a tank. The most important characteristic of these devices is the invariable quantity of solution delivered during each time interval. Many so-called feeders (such as the flow from a restricted faucet or through an orifice plate attached to a container of liquid where the head on the orifice varies) are entirely unsuitable for fluoride feeding because of the wide variations in flow.

There are essentially four different types of satisfactory solution feeders available: diaphragm, piston, rotating cup, and decanting

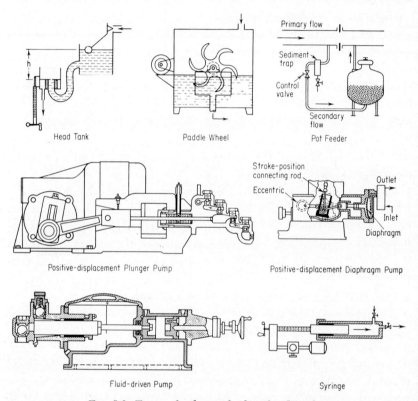

FIG. 8.2. Types of solution feeders for fluorides.

arm. The operating principles of all these are illustrated in Figs. 8.2 and 8.6.

The most widely used solution feeders in the waterworks industry are the diaphragm types. The diaphragm, made of flexible rubber, plastic, or thin metals, is actuated (as shown in Fig. 8.3) either by a reciprocating plunger attached directly to the diaphragm or indirectly by means of varying the pressure periodically on a confined volume of hydraulic fluid. The reciprocating motion of the piston is obtained from either a slowly rotating cam or a crank.

Variation in the capacity or the rate of feeding a solution is obtained by changing the speed of operation (depending on the motor speed, relative sizes of driven and driver pulleys, arrangement of gear reducers, and the like) and by varying the length of the stroke. The designs for varying the length of stroke differ widely and in many cases are the distinguishing characteristic of a particular feeder.

Diaphragm feeders are ideally suited for medium-pressure service —up to about 125 psi (pounds per square inch). They should not be used, however, against pressures less than about 5 psi and particularly should never be used against a vacuum, such as that obtained in the suction side of a pump. A constant positive pressure on the discharge is a guarantee of their continued accuracy. Some feeders are equipped with spring- or rubber-loaded discharge valves that assure the maintenance of such positive pressures. Negative suction heads should not ordinarily exceed about 4 ft and should remain constant.

Fig. 8.3. Diaphragm-type solution feeder. (*B-I-F Industries, Inc.*)

Diaphragm and piston-type feeders are driven by almost any source of power which can impart a reciprocating motion to the piston. These include electric motors of various standard speeds, belt drives from shafts driving water pumps, and other rotating machinery in a water plant; gasoline motors; hydraulic or pneumatic pistons, with the energy obtained either from an auxiliary water pump or air compressor or from the pressure in the line carrying the water to be treated; or by a solenoid periodically energized by a timer or contactor actuated by a water meter or other pacing mechanism. The principal characteristic of such prime movers is that they are operated at a constant speed to produce a uniform solution delivery or at a speed proportional to the quantity of water to be treated or at the rate required by a fluoride transducer (see Chap. 9).

The piston or plunger pump (Fig. 8.4) is similar to the diaphragm type except that the piston directly displaces the quantity of solution to be fed. By reducing the diameter of the piston or the length of the stroke, very minute quantities of solution can be accurately fed. The length of the stroke is usually controlled by an adjustment of the crank arm or by limiting the length of contact against a rotating cam. Such adjustments can usually be made with the feeder running. As shown in Table 8.1, some plunger feeders can deliver accurately as little as 0.375 ml per min. This type of feeder is driven by the same means as outlined for diaphragm feeders.

One disadvantage of the piston type is the presence of a stuffing box, which must be designed to resist the pressure of the solution

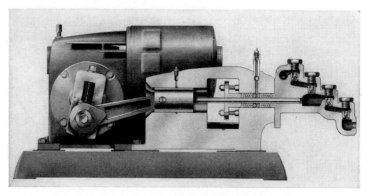

Fig. 8.4. Piston-type solution feeder. (*Milton Roy.*)

being fed so that no objectionable leaks occur around the piston. In a diaphragm feeder there is no chemical packing or possibility of leaking through a packing gland. On the other hand, the permissible operating pressure for piston types is considerably higher than for diaphragm feeders, much beyond the range of any pressures encountered in waterworks practice. Some such feeders are designed for pressures up to 10,000 psi.

In both types of feeders a pulsating flow will occur because of the reciprocating nature of the operating mechanism. Ordinarily, this is not objectionable if variations in fluoride levels cannot be detected in the distribution system. If, however, the closest consumer is served with water which shows a fluoride level varying more than 0.1 ppm during a feeder-stroking cycle, then means should be provided to suppress this variation. This can be done in three ways. (1) A mixing basin or detention tank can be inserted in the line after the point of application of the fluoride solution. If such a tank is large enough, the fluoride solution should blend completely with the water. (2) The frequency of feeder stroking can be increased with a proportional weakening (dilution) of the fluoride solution. This will also require a proportional reduction in the amount of solution delivered per stroke. (3) Dual feeders can be used so that the feeding stroke of one will occur during the intake stroke of the other. This plan also requires a dilution of the fluoride solution.

The delivery range of a particular piston or diaphragm feeder is usually limited between a fraction of its maximum delivery up to the maximum delivery at a particular maximum rotating speed, the minimum usually being between one-sixth and one-sixtieth of the maximum. In other words, if the maximum range is in the order of 10 to 1 (1/10) and the maximum delivery is 100 ml per min, then the minimum rate should not be less than about 10 ml per min. Although most feeder manufacturers provide means for reducing the minimum rate of delivery below the recommended rate, accuracy is markedly reduced in the lowest range. This reduction in accuracy is caused by check valves which either will not open or will leak at the lower rates. On the other hand, extremely high operating (rotating) speeds also markedly affect accuracy because insufficient time is provided for the cylinder in the piston-type feeder or the space formed by the diaphragm to fill completely on the suction stroke.

For very small rates of delivery a modified plunger-type feeder is available. The plunger is withdrawn, and the solution flows into the chamber. A screw-type drive slowly moves the piston forward, discharging the solution in a manner similar to the action of a hypodermic syringe. Two of these feeders can be driven in tandem, one filling while the other discharges. Such feeders are available from the American Instrument Company, Inc., Silver Spring, Maryland, and the Omega Machine Company, Providence, Rhode Island.

Neither the head tank nor the pot feeder (shown in Fig. 8.2) is used in this country for fluoridation because of their inaccuracy, which is caused by clogging of the orifices, variations in the solubility of the chemical to be fed, and channeling of the water through the bed of chemical. They are, however, being used experimentally for this purpose in several foreign countries (Japan and Brazil) where constant attention is routinely available.

The paddle wheel (rotary dipper) (also shown in Fig. 8.5) is widely used for accurate solution feeding. A constant-head device maintains the level of solution in a tank from which a series of cups attached to a partially submerged wheel lifts and discharges a constant volume of solution. Rate-of-feed variations are obtained by (1) changing the size of the dippers or cups, (2) adding or removing cups from the rotating wheel (3) changing the speed of rotation, or (4) varying the proportion of the contents discharged from the full cups, the rest being returned to the tank. (A cutoff device is adjusted so

Fig. 8.5. Rotary-dipper solution feeder. (*B-I-F Industries, Inc.*)

that a portion or all of each full cup is discharged after each cup has emptied.) This type of feeder is intended for those installations in which the solution can be fed into another tank or basin or into a channel below the level of the feeder. It is not intended for feeding against any pressure. These feeders are manufactured by:

Denver Equipment Company, Denver, Colorado
Infilco, Inc., Tucson, Arizona
Omega Machine Company, Providence, Rhode Island

The ranges of feeds for the various makes are: Denver Equipment Company, up to 120 gal per hr; Infilco, Inc., up to 500 gal per hr; Omega Machine Company, up to 1,800 gal per hr (range 100 to 1).

Feeder Types and Capacities

In another type of solution feeder the operation is based on a decantation principle, whereby a thin upper layer of the solution in a tank is constantly being removed (Fig. 8.6). This is done by gradually lowering an open-ended pipe or hose into the solution to be fed. The other end of the hose or swing pipe is connected to a fitting on the lower side of the tank which permits the solution to drain from the

FIG. 8.6. Decantation-type solution feeder. (*Permutit.*)

hose to a point of application outside the tank. The open end of the hose or swing pipe is lowered by means of a clockwork mechanism (either electric or spring-driven). Variations in delivery up to 100 to 1 are obtained by changing the output speed of the gear or ratchet speed reducer on the clock. Tanks are of various sizes, but the usual tanks hold about a 1-day supply of solution. Rates of discharge in the standard feeders range from 2.5 ml per min to over 2 liters per min (32 gal per hr) at an accuracy of 3 per cent. A model complete with an 18-gal tank costs $500.

As in the case of the rotating-dipper feeders, the decantation type cannot be used to feed against any pressure above atmospheric. However, these feeders can be made to discharge into a pump suction box, from which the solution can be pumped to higher-pressure areas. One such arrangement is shown in Fig. 10.1.

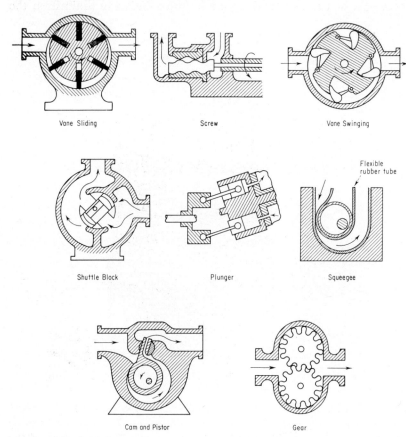

FIG. 8.7. Rotary-displacement-type solution feeds. (*USPHS.*)

The decantation feeders are manufactured by the Graver Water Conditioning Company, New York, New York, and the Permutit Company, Inc., New York, New York.

The drawings in Fig. 8.7 represent a group of rotary volumetric, positive displacement solution feeders which are used primarily where higher capacities are required. The ranges of some, in fact, are so high that they are at present seldom used in fluoridation practice.

Where the capacities of such feeders might be used, it would normally be more economical to use a more insoluble, cheaper compound applied with a dry feeder. In addition, these feeders are generally somewhat more complicated and difficult to manufacture and are consequently more expensive. They are all capable of being adjusted to control of rate of delivery, which is done by changing the speed of rotation either by means of changing the speed of the motor or by means of gear or belt speed reducers.

The variation in the cost of solution feeders, which ranges between $150 and several thousand dollars, is due to size (rate of delivery), pressure capabilities, ease and range of adjustment, accessories (for instance, a stainless-steel or fiberglass sodium fluoride saturator tank costs about $500), and refinements in manufacture. In general, diaphragm feeders of the type ordinarily used in waterworks practice cost between $350 and $450, including the driving motor and plastic solution lines. With the usually furnished plastic reagent heads and rubber diaphragms, there should be no difficulties on account of the corrosiveness of the fluoride solutions, including hydrofluosilicic acid.

DRY FEEDERS

Dry feeders are those devices which deliver a measured quantity of dry chemical during a particular interval of time. There are two basic types: volumetric feeders, which deliver a specific volume (cubic inches, cubic feet, quarts, pecks, bushels, barrels, and similar volumetric designations) of chemical material during a set time; and gravimetric feeders, which deliver a certain weight (ounces, pounds, grams, and the like) during a selected time interval. Generally volumetric feeders can deliver smaller quantities than gravimetric feeders, but the principal differences in performance are: (1) volumetric feeders have an accuracy of from 3 to 5 per cent by weight as compared with 1 per cent for gravimetric feeders; (2) volumetric feeders are simpler and of less expensive construction; (3) gravimetric feeders are generally more readily adapted for recording the quantities of compounds fed and for automatic control. Gravimetric feeders are almost invariably used in the larger plants.

Many types of volumetric feeders can be converted to gravimetric types by attaching a weighing mechanism to the entire assembly. The delivery from such a feeder is then automatically controlled by the loss in weight resulting from the removal of some of the compound from the hopper. One of these is shown in Fig. 8.8.

Volumetric feeders are essentially a combination of a driving mechanism, a means for delivering a constant volume of dry compound, a hopper for holding the compound, and a chamber for dissolving the compound before discharge into the water supply.

FIG. 8.8. Volumetric feeder converted to gravimetric operation. (*Wallace & Tiernan.*)

The driving mechanism is almost invariably an electric motor, the speed of which has been reduced through gears or belt drives; occasionally belts from other rotating equipment, such as from the shaft of a pump, are used as the driving mechanism.

The chemical-delivery mechanism is a convenient means for distinguishing one type of volumetric feeder from another. Almost every manufacturer has a different design for feeding chemicals volumetrically. These might be classified according to several types: rotating

Feeder Types and Capacities

disk, oscillating pan, vibratory pan, reciprocating screw, rotating roller, star wheel, and combinations of these principles.

Brief descriptions of the various types are given here merely to indicate broadly the principles of operation. Much more complete details are readily obtainable from the manufacturers.

The rotating-disk feeders are probably the most widely used, particularly in the smaller plants. They are fitted with a horizontally rotating flat steel disk. The chemical to be fed flows from the hopper onto the center of the disk and is supported by the rotating disk. A scraper or orifice removes a portion of the material as the disk slowly rotates. For very low rates, a groove is cut in the face of the disk and the material is removed from the groove by an appropriately shaped scraper. The widths of the grooves can be varied for different rates of delivery. Feed rates can also be changed by controlling the speed of rotation of the disk and by adjusting the position of the scraper to govern the amount of material removed. Rates can be as little as 1 oz sodium fluoride per hr (75 lb per cu ft) up to 400 lb per hr. This type of feeder is made by:

Fig. 8.9. Volumetric disk dry feeder on scales. (*B-I-F Industries, Inc.*)

Denver Equipment Company, Denver, Colorado
Infilco, Inc., Tucson, Arizona
Omega Machine Company, Providence, Rhode Island

A typical machine is shown in Fig. 8.9, which also shows how the entire machine is mounted on scales used to check the quantity of material delivered, and how the bags of compound are emptied so that a minimum amount of dust is created. A typical disk feeder with a maximum capacity of 11 lb sodium fluoride per hr and equipped with an 11-gal dissolving chamber and a 3¼-cu-ft hopper costs about $1,000.

The oscillating-pan (oscillating hopper) type of feeder consists essentially of a flat narrow pan or trough into which the fluoride compounds fall from a hopper above. Either the pan or the lower part of the hopper slowly oscillates along the axis of the pan, forcing the removal along the two open edges of the pan of a portion of the chemical in the pan. A cross section of the mechanism utilizing the oscillation of the hopper bottom is shown in Fig. 8.10. Delivery rates are controlled by both the speed of oscillation and the length of the

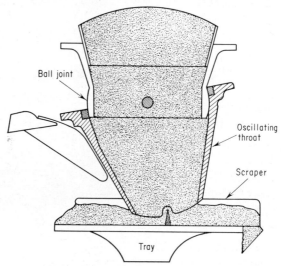

Fig. 8.10. Oscillating-pan volumetric feeder. (*B-I-F Industries, Inc.*)

oscillary stroke or the thickness of the chemical on the pan. The various sizes depend on the delivery ranges—from 2 oz per hr up to 500 lb per hr of sodium fluoride (75 lb per cu ft) with a range within a particular feeder of 40 to 1. This type of feeder, with a 3¼-cu-ft hopper and a 10-gal dissolving chamber, costs $1,040. It is manufactured by:

F. B. Leopold Company, Inc., Pittsburgh, Pennsylvania
Infilco, Inc., Tucson, Arizona
Omega Machine Company, Providence, Rhode Island

The vibratory-pan feeder is a device for discharging a volume of chemical from a pan, chute, or trough made to vibrate electrically. A magnet is energized by means of a pulsating current (either ordinary alternating current or rectified, pulsating direct current). The trough is mounted on springs and connected directly to the magnet as shown in Fig. 8.11. The action of the tray is downward and backward on the power stroke and upward and forward on the next stroke through the action of the springs. The material on the tray moves forward slightly on each stroke and appears to flow like water because of the

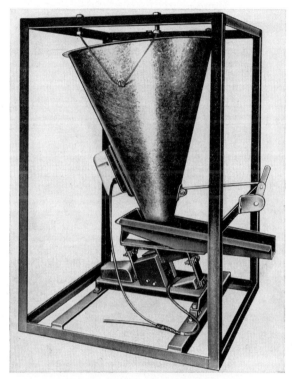

FIG. 8.11. Vibratory-pan volumetric feeder. (*Syntron Co.*)

high stroking frequency (3,600 strokes per min on 60-cycle current). The rate of delivery is controlled by a rheostat, which determines the voltage and consequently the degree of movement of the trough. A wide range of sizes is available, the range of delivery being from 2 g per hr to 8,000 lb or more per hr. In water plants, however, only the smaller units might be used for fluoridation; the larger plants obtain much better accuracy with gravimetric feeders. The discharge from vibratory feeders can be utilized on belt-gravimetric-type feeders, which will be described later. The smallest feeders of this type capable of delivering up to a maximum of 500 lb of material per hr costs $80 with controls but without a hopper or other accessories. Power consumption of this feeder is only 3 watts. Feeders of this type are available from:

Eriez Manufacturing Company, Erie, Pennsylvania
Jeffrey Manufacturing Company, Columbus, Ohio
Syntron Company, Homer City, Pennsylvania

A feeder which combines this vibratory principle and a continuously rotating-screw conveyor is available. The vibrations are imparted to the entire feeder (with the exception of the hopper), which completely fills each flight of a spiral conveyor. Ranges in sizes provide deliveries from 1 oz to 100 tons sodium fluoride per hr. The cost of the smallest of these feeders is $645 complete. They are made by Vibra Screw, Belleville, New Jersey.

Another type of rotating-screw-conveyor feeder is made by the B. F. Gump Company, Chicago, Illinois. This feeder consists of a slowly rotating helical conveyor. The material to be fed floods the conveyor from a hopper (shown in Fig. 8.12) and is discharged from a trough at the end of the screw. The spiral shaft is rotated by either a gear train or a ratchet mechanism. Rate of feed is controlled by the length of sweep of the ratchet and also by selecting the gear-reduction ratio of the drive shaft. Range is from $\frac{1}{5}$ cu ft per hr (17 lb sodium fluoride per hr) and upward. This feeder could probably be used to feed only the crystalline grade of chemical (sodium fluoride) in the smaller plants because other more hydroscopic chemicals tend to clog it. The cost of the smallest models is $450.

Another adaptation of the screw-conveyor feeder is shown in Fig. 8.8. In this model (made by Wallace & Tiernan, Inc., Belleville, New Jersey) the slowly turning screw also has a forward and backward motion. This action occurs within a tube from the end of which the chemical is discharged. The range of feed within a particular size of feeder is in the area of 120 to 1. Several different sizes are available, providing a broad capacity selection: from less than 1 lb per hr to over 170 lb per hr. Adjustment in rate of feed is obtained by controlling operating speed of the driving mechanism together with the length of stroke of the oscillating-screw conveyor.

Feeder Types and Capacities

Fig. 8.12. Rotating-screw dry feeder. (*B. F. Gump.*)

A rotary feeder known variously as "star wheel" or "rotolock" can be used as a volumetric feeder, although these devices are commonly incorporated in gravimetric feeders. Fig. 8.13 shows a common design whereby a spoked wheel rotating slowly within a snug-fitting housing delivers a constant volume of material from within each vane

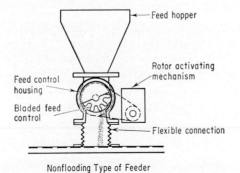

Fig. 8.13. Star-wheel volumetric feeder. (*B-I-F Industries, Inc.*)

of the wheel. A hopper is fitted on the upper end and the dissolving chamber on the lower. Wheels of many designs are available: diameter, number of spokes, length, and rotational speed are a few of the different characteristics designed for different quantities and kinds of material to be handled. This type of feeder is almost proof against flooding; i.e., that characteristic of some feeders to permit the chemical to flow like water through the metering elements. Flooding permits a large quantity of chemical to flow through the feeding mechanism in a very short time, clogging the dissolving chamber and

FIG. 8.14. Feed-roll volumetric feeder. (*Wallace & Tiernan.*)

causing a large overdose. On this type of feeder a positive means is provided to prevent this. Such feeders are almost always used in the largest installations, where delivery is controlled gravimetrically. Variations in feed are accomplished by choice of rotor and its speed or frequency of operation. These feeders may be made as large as desired, some being available for up to 10 tons per hr. Equipped with a $4\frac{1}{4}$-cu-ft hopper and 30-gal dissolving chamber they cost $1,700. They are made by:

Allen-Sherman-Hoff Company, Wynnewood, Pennsylvania
Omega Machine Company, Providence, Rhode Island

Prater Pulverizer Company, Chicago, Illinois
Pulverizing Machinery Division, Metals Disintegrating Company, Summit, New Jersey

Another rotary type of volumetric feeder for fluoridation was developed from a device used to feed a bleaching compound into flour. Here a thin ribbon of chemical is extruded between two slowly rotating rollers, similar in action to a clothes wringer. This type of feeder is shown in Fig. 8.14. Feed rate can be altered by changing either the width of the ribbon of chemical (by blocking access to variable portions of the rollers) or the speed of rotation of the rollers. The range of feeding is up to 20 lb per hr and down to 0.13 lb per hr of sodium

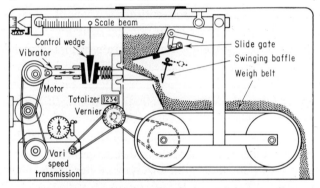

FIG. 8.15. Moving-belt gravimetric feeder. (*Permutit.*)

silicofluoride (65 lb per cu ft). The cost is $1,025 including the 0.6-cu-ft hopper and a 25-gal dissolving chamber. It is made only by Wallace & Tiernan, Inc., Belleville, New Jersey.

A moving belt is the principal characteristic of another group of volumetric feeders. As shown in Fig. 8.15, the belt moves slowly past the lower opening of the hopper, where it is loaded to a constant level and width of chemical. At the end of the belt travel, the chemical is discharged to the dissolving chamber. Rates are changed by controlling the speed of the belt. The minimum rate of delivery is about 0.6 lb sodium fluoride per hr with possible rates of up to 2 tons per hr in the largest models. Accuracy is in the order of 2 per cent. The smaller of two models costs $1,500 with a $4\frac{1}{2}$-cu-ft hopper and a 50-gal dissolving chamber. These feeders are manufactured by the Permutit Company, Inc., New York, New York.

GRAVIMETRIC FEEDERS

Gravimetric feeders feed chemicals at a constant weight rather than at a constant volume during a given period of time. Because of their great accuracy they should be used for feeding fluoride compounds wherever possible. Only when the minimum demand for chemicals is less than about 10 lb per hr should other types of feeders be considered. Gravimetric feeders are also quite readily adaptable for recording quantities of chemical feed and for automatic control. Another advantage is that a constant weight will be fed even if the bulk density of the compound has changed.

The weight of the chemical must be continuously taken because control of these feeders is based entirely on weight. Such control is accomplished in two basic ways. In one the container (storage bin or hopper) is continuously weighed and the rate of loss in weight of the material in the hopper is automatically maintained by prior selection of the rate of feed. The discharge is so regulated that the material left in the hopper follows a linear reduction in weight. In the other type, a section of a moving belt carrying the compound is continuously weighed. The flow of compound onto the belt is controlled by the deviations from a desired preset rate of discharge.

The first type (loss-in-weight), shown in Fig. 8.16, consists of a hopper suspended from a scale system, an electrical-mechanical system for moving the poise on the scale beam, a mechanical means for moving the compound from the hopper in an amount depending on the position of the scale beam, and a dissolving chamber. The lead screw drive (a synchronous motor) moves the poise along the beam at a preset rate of speed. If more material is momentarily fed than indicated by the position of the poise, then the beam will lower. This action moves the control wedge (near the oscillator) downward, permitting a decrease in the amplitude of the stroke driving the star wheel or vibratory feeder mechanism. Less material will then be delivered until the weight of compound remaining in the hopper is again balanced by the weight on the scale beam. The error in feeding in this type of feeder is generally less than 1 per cent. The minimum delivery is 10 lb per hr with a range of feed in the order of 100 to 1, while some models can deliver more than 2 tons per hr. The smallest size, with a 5-cu-ft hopper and a 40-gal dissolving chamber, costs $2,500.

Feeder Types and Capacities   113

The other type of gravimetric feeder, in which a section of a loaded, moving belt is continuously weighed, is shown in Fig. 8.15. The weight of the belt is balanced by a scale beam, the position of which controls delivery of the compound onto the belt. Any deviation from this weight on the belt causes more or less material to fall onto

Fig. 8.16. Gravimetric dry feeder. (*B-I-F Industries, Inc.*)

it from the hopper. While a mechanical vibrator is shown in the drawing for moving material from the hopper to the belt, other methods are available. On this model, vibrations imparted to a diaphragm on the hopper are generated by an eccentric and transmitted through a wedge which varies the amplitude of the vibrations, depending on the position of the scale beam. Other means for loading the belt include mechanically, electrically, or pneumatically actuated adjustable gates on the hopper, electric vibrations on a feed pan, or a rotating star wheel. Accuracy in these feeders is also in the order of 1 per cent or less. Range of feed is as much as 100 to 1, and adjustments are readily made merely by moving the poise on the scale beam. A feeder of this type delivers up to 200 lb per hr; it costs $2,000 without accessory equipment.

It is apparent that almost any volumetric feeder (including the vibratory types) can be adapted to perform as a gravimetric feeder either by placing the entire volumetric feeder on a scale and controlling the discharge according to the predetermined loss in weight, or by weighing the loaded belt section and controlling the discharge from a hopper by deviations from the preset rate. However, only those gravimetric arrangements described are at present available. The following are manufacturers of gravimetric feeders:

B-I-F Industries, Providence, Rhode Island
Infilco, Inc., Tucson, Arizona
Syntron Company, Homer City, Pennsylvania
Vibra Screw Feeders, Inc., Clifton, New Jersey
Wallace & Tiernan, Inc., Belleville, New Jersey

Feeder Accuracy

Fluorides should be fed more accurately than any other material used in treating water. The reason for this is apparent when the results shown in Chap. 2 are considered. From the point of view of optimum reduction in dental decay, a difference of only 0.3 ppm fluoride below the optimum level will on the average produce about one additional decayed tooth among 13-year-old children. On the other hand, as shown in Chap. 1, prolonged use of water containing fluorides in excess of the optimum will produce mottling (dental fluorosis). After the optimum level is determined (see Chap. 7), deviations greater than 0.1 ppm (10 per cent when the optimum level is other than 1.0 ppm) should not be tolerated. A considerable amount of evidence shows that when prolonged deviations occur, the effect on children's dental health is apparent. For instance, where the fluoride concentrations have been maintained at variable and consistently lower levels than the optimum, the DMF rates have been much higher than in similar communities where the level has been maintained correctly.[1]

[1] J. E. Chrietzberg and F. D. Lewis, Jr., Effect of Inadequate Fluorides in Public Water Supply on Dental Caries, *J. Georgia Dental Assoc.*, **31**: 10–14 (1957).

A standard of 0.1 ppm deviation from 1.0 ppm is in fact relatively easy to obtain after some experience and training of waterworks operators and with the feeding equipment described heretofore. A deviation of 0.1 ppm at 1.0 ppm level is an error of 10 per cent. Few feeders today are this much in error; some, in fact, have errors of less than 1 per cent. Nevertheless, feeders of the utmost accuracy should be sought in every case because not only do all feeders have a certain inaccuracy, but there may be deviations in the purity of the chemical used, or the pacing and control equipment may contribute errors, or manual adjustment of the feeder delivery may be in error or too long delayed.

The accuracy of a feeder is usually designated as the percentage deviation from an established rate. At least two inconsistencies arise from this practice. The first derives from a definition of "established rate." This may mean the rate as indicated on an adjustment scale of the feeder. If a feeder is set to deliver, say, 10 lb per hr and tests indicate a uniform delivery of 11 lb per hr, the error is 10 per cent. This error, however, is not serious, inasmuch as it has probably been caused by a deviation in positioning the scale plate and can be readily corrected by lowering the setting to the desired rate of feed. On the other hand, a feeding rate can be set and tests will indicate that the rate of delivery is the one desired. Subsequent tests may show a deviation from this rate. This error is entirely different from the one above and is more serious. The latter figure should be used to designate the accuracy of the machine.

In designating the accuracy of a feeder, the range of rate or the set rate of feed should be described which corresponds to the percentage error or "repeatability" figure given (plus or minus). Errors generally increase at the extreme upper and lower ranges of a feeder. In addition, the time interval applicable to this accuracy should be given. This interval is far from being standard among different manufacturers, but generally one-half hour should be sufficient for collecting a representative volume. Too short an interval may introduce an error which is compensable in a longer test. Prolonged tests provide no additional information.

Solution feeders are tested against the pressure to be encountered by measuring the discharge in a graduate or suitable measuring container. Errors in solution feeders may be caused by belt slippage, voltage variations, gland leakage, diaphragm wear, or improperly seated valves. A means for checking the delivery of solution feeders without removing the suction or discharge piping is shown in Fig. 8.17.

Dry feeders are tested by collecting the material delivered in a shallow pan held over the dissolver and weighing the collected material. Errors in dry volumetric feeders occur primarily because of the compressibility or classification of the material being fed. Fluffy grades of powdered fluorides will be more compressed when the hopper is full than when nearly empty. Material containing particles of a considerable size range may be classified in handling, in shipment, or from vibrations transmitted to the feeder hopper. Resulting changes in density can produce errors amounting to as much as 10 per cent.

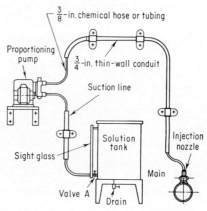

Fig. 8.17. Means for checking accuracy of solution feeders. (*B-I-F Industries, Inc.*)

Such errors are entirely eliminated in gravimetric-type feeders, in which variations seldom exceed ±2 per cent.

By combining the probable errors of the several components of a fluoride feeding system, its overall error can be computed. For instance, referring to Table 8.2, it is possible to calculate the probable error of a system composed of a gravimetric dry feeder (1 per cent error) controlled by means of an impulse-duration system (0.75 per cent error). The probable error would then be the square root of the sum of the squares of these errors, or 1.25 per cent.

Despite the great accuracy of most feeding systems, errors in the fluoride level will result if the system is maintained or operated in a manner which does not contribute to its optimum performance. This has in fact happened at some water plants, and, in some very few

TABLE 8.2. ERRORS FOR DRY AND LIQUID FEEDER PACING
AT 100 PER CENT OF FLOW†

Item	Impulse duration (per cent pulse)		Pneumatic (3–15 psi)	
	Maximum, per cent	Probable, per cent	Maximum, per cent	Probable, per cent
Dry feeder, volumetric..........	5.75	5.05	6.50	5.12
Dry feeder, gravimetric..........	1.75	1.25	2.50	1.50
Metering pump................	2.75	2.13	3.50	2.29
Feedback ratio control (liquids)..	....	....	2.00	1.41

Individual accuracies, per cent

Impulse duration.........	0.75	Dry feeder (volumetric).....	5.0
Pneumatic.............	0.5	Dry feeder (gravimetric)....	1.0
Vacuum...............	1.0	Metering pump...........	2.0
		Chlorinator..............	4.0
		Feedback loops...........	1.0

Maximum error	*Probable error*
$= E_1 + E_2 + E_3 + \cdots E_n$	$= \sqrt{E_1{}^2 + E_2{}^2 + E_3{}^2 + \cdots E_n{}^2}$

† Data from the Foxboro Co.

places, such errors have been considerable and consistent. At most places, however, the errors have been very small. Figs. 8.18 and 8.19 show how accurately the fluoride concentrations can be made to conform to the predetermined optimum level. In the case of Grand

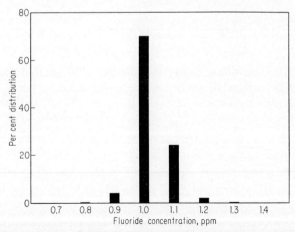

FIG. 8.18. Consistency in fluoride levels at Grand Rapids. [G. C. S. Spitz, F. B. Taylor, and W. L. Harris, Experience in Maintaining Constant Fluoride Concentrations, Am. J. Public Health, 48(12): 1651 (1958).]

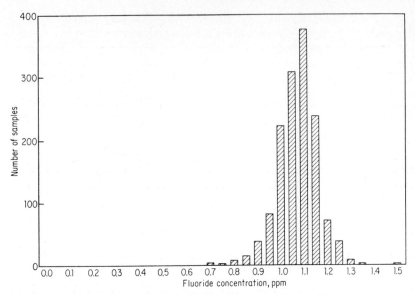

FIG. 8.19. Consistency in fluoride levels at Newburgh. (*USPHS.*)

Rapids, the aim was to maintain a fluoride level between 1.0 and 1.1 ppm; at Newburgh, between 1.0 and 1.2 ppm. In both cases, the vast majority of the samples fell within these limits.

DISSOLVING CHAMBERS

Dissolving chambers are a part of all dry feeders and, if properly designed, should dissolve continuously all chemicals fed into them before being discharged. Some few installations have attempted to discharge the chemicals from dry feeders directly into clear wells, mixing basins, and other points in the water plant, generally with indifferent results. In many cases the chemical falls directly to the bottom of the basins and remains there undissolved, piling up until it must be removed. In almost every case a dissolver should be included with every dry feeder.

If any chemical leaves the dissolving chamber undissolved (discharged as a slurry), it indicates either (1) that the dissolving chamber is too small, (2) that the detention time is too short, (3) that too little water is being provided, (4) that insufficient mixing is being provided, or (5) that short-circuiting is occurring and permitting some of the chemical to leave the dissolver almost immediately after being introduced.

The tables of solubilities of chemical compounds are obtained under ideal conditions of time, temperature, and purity of solvent (water). In order to obtain continuously a solution of a dissolved fluoride compound from a dissolving chamber, it is generally advisable to limit the maximum solution strength to one-fourth of that of a saturated solution at ordinarily encountered water temperatures. This is based on a theoretical minimum detention time of 5 min in the dissolving chamber. On this basis, sodium fluoride, for instance, is dissolved to make solutions of about 1 per cent strength. A 1 per cent solution is formed by dissolving 1 lb chemical in about 12 gal water. If it is assumed that 10 lb sodium fluoride per hr is fed, then 120 gal water will have to be provided each hour, or 2 gpm. With a 5-min detention time, the dissolving chamber will have a capacity of at least 10 gal. From these figures it is apparent that for each pound of sodium fluoride fed per hour we would need one gallon of dissolving-chamber capacity. Other fluoride compounds, which are listed in Table 7.3 and which might be used in dry feeders, are computed similarly, as shown in Table 8.3.

TABLE 8.3. DISSOLVING-CHAMBER CAPACITIES REQUIRED FOR VARIOUS FLUORIDE COMPOUNDS

Fluoride compound	Maximum practical solution strength, %	Gal water required per lb of chemical fed to provide practical solution strength	Dissolving-chamber capacity (gal) per lb of chemical fed with 5-min detention
Sodium fluoride	1.0	12	1.0
Calcium fluoride	Will not dissolve in water; must be used as described in Chap. 7		
Sodium silicofluoride	0.2	60	5.0
Magnesium silicofluoride	15	0.8	0.07
Ammonium silicofluoride	5	2.4	0.2
Potassium fluoride	25	0.5	0.04

In practice dissolving chambers are seldom made with less than 5-gal capacity. This is because (1) it is not economical to make smaller ones, (2) chambers of at least this size are required to support the smallest feeders, and (3) short-circuiting is more likely to occur in the smaller tanks.

Mixing of the chemical and the water is accomplished with either water jets or electric mixers. The jets are operated by the pressure of the incoming water, and in some feeders this pressure is used to rotate a paddle mixer (Fig. 8.20). Electric mixers are sometimes required in fluoride feeder dissolvers primarily because ample detention times have not been provided.

Fig. 8.20. Rotary paddle mixer in dissolving chamber. (*Wallace & Tiernan.*)

Dissolving chambers should be so baffled that short-circuiting is reduced to a minimum (when the least amount of undissolved material is discharged from the chamber). Baffling should be so designed that the path of the chemical to the outlet of the chamber requires the theoretical retention time for completion.

Once the size of the dissolving chamber is fixed, the detention time will depend entirely on the rate at which water is introduced. The rate should be fixed as accurately as possible; too much water will reduce the detention time and thereby reduce the opportunity for the chemical to dissolve; too little will increase the solution strength and thereby reduce the speed at which the compound dissolves. As the solution strength approaches saturation, it may require weeks or even months for some fluoride compounds to dissolve completely.

For attaining accurate control of the water entering the dissolving chamber, various meters and gauges are available. A water meter will show the total quantity used over a certain time interval and provide a guide to overall water use. Flow-indicating devices, such as those made by the Brookes Rotometer Company, Lansdale, Pennsylvania, Fischer & Porter Company, Hatboro, Pennsylvania, and Schutte & Koerting Company, Cornwells Heights, Pennsylvania, are valuable aids in adjusting water flow. The water line may also include a solenoid valve (to shut off and turn on the water automatically when the feeder is stopped or started), a pressure regulator (to ensure a constant water flow), and a needle valve (to regulate the flow).

Some thought has been given to increasing solubility and dissolving time of various chemicals by increasing the temperature of the water. This is a definite possibility because many chemicals, including sodium silicofluoride (see Fig. 7.5), have a higher solubility as the temperature of the solvent increases. However, this method is seldom used (except in Alaska) because the rising vapors from the dissolving chamber moisten the chemicals in the hopper and cause them to clump together and to adhere to the feeding mechanism. Both of these effects seriously impair the accuracy of feeders.

Many other available items of auxiliary equipment are desirable or necessary in particular installations. They will be described in Chap. 9.

PLANNING THE BEST SYSTEM

With some knowledge of the availability and characteristics of fluoride compounds (Chap. 7) and the many different kinds of chemical feeders, it is possible to select the best fluoride feeding system. The design of such a system should be based on a consideration of three principles: (1) reliable performance, (2) utmost safety, and (3) the most economical equipment. The engineer must choose the combination of the most economical chemical and appropriate feeding system commensurate with the size and facilities (personnel and type of plant) of the community.

In order to estimate the size and other characteristics of the various pieces of equipment, it must first be determined how much fluoride compound will have to be fed in successively larger increments of time. The quantity of water to be treated must be measured at various intervals during different times of the day, seasons of the year, or other periods affecting the maximum consumption of water. For sizing the feeder itself, the basic consideration is to know the quantity of compound required per unit of time (a minute being most convenient) which will be added to a given quantity (for computation purposes, the maximum quantity) of water to be treated. The quantities of water delivered during intervals longer than a minute (intervals of up to a year) are used for sizing other parts of the fluoridation system.

The amount of compound used is directly proportional to the quantity of water to be treated. Consequently it must be determined how much water will be treated during the period of maximum rate of flow. This rate can be determined by several means:

1. Water-meter readings
2. Pump discharge ratings
3. Measurement of rise or fall of water level in a basin or tank during an interval of time
4. Use of velocity or flow-measuring devices—propeller meters, pitot tubes, pressure or head differentials derived from orifice plates, flumes, weirs, and flow tubes

Whatever the means used, care must be taken to assure that the reading eventually used is actually the maximum that is likely to be encountered. There are occasionally short periods during which the rate of water consumption is unusually high. These might be the result of fires, main breaks, or other rare or unusual circumstances. For the purpose of designing for fluoridation equipment they can usually be ignored, because such infrequent intervals (for example one day per year) of low-fluoride water use will have no measurable effect on reducing tooth decay.

The maximum rate at which water must be treated during the period of a day or so (rather than during a minute) should also be known in order to determine the size of the chemical hopper on the feeder. Generally, in the smaller plants a hopper sufficient to store a day's supply of chemical is sufficiently large. Similarly the maximum rate of flow over a period of a week or two should be known in order

Feeder Types and Capacities

to compute the proper size of storage bins or areas. The interval selected for this purpose will, of course, depend on how quickly chemicals can be delivered to the plant after an order has been placed. In addition, the maximum water used during a year should be estimated in order that budgets can be devised for the purchase of chemicals. In many cases it will be found economical to purchase yearly quantities of chemicals on a single order even though the whole amount cannot be stored at the plant. Many vendors, in fact, prefer to deliver chemicals at specified intervals in smaller quantities rather than delivering a full year's supply at once.

The following computation is included in order to illustrate how a feeder might be selected:

Assume that

1. Maximum rate of water delivery at the point of fluoride application is   575 gpm
2. Optimum fluoride concentration desired is   0.90 ppm (see Chap. 6)
3. Least fluoride content ever found in untreated water is   0.15 ppm (see Chap. 6)
4. Maximum quantity of fluorides to be fed is   0.75 ppm (difference between lines 2 and 3)

Since 1.0 ppm is equal to 8.3 lb per million gal (see Chap. 1, page 4, for definition and explanation of this ratio), then:

5. 0.75 ppm requires $0.75 \times 8.3 = 6.25$ lb fluoride ion per million gal water

and:

6. At the 575-gpm rate, $\dfrac{1 \text{ million gal}}{575 \text{ gpm}} = 1{,}750$ min are required to pump 1 million gal; then:

7. $\dfrac{6.25 \text{ lb} \times 16 \text{ oz per lb}}{1{,}750} = 0.0577$ oz per min of fluoride ion is required (or 1.63 g per min)

If, for example, this quantity of fluoride ion would be obtained from sodium fluoride, then we would have to supply $0.0577 \times 2.21 = 0.1275$ oz of 100 per cent pure sodium fluoride per min or 7.6 oz per hr. The figure 2.21 is the amount (ounces) of sodium fluoride required to obtain each ounce of fluoride ion. It is computed by dividing the combining weight of sodium fluoride (42) by the equivalent weight of fluoride ion (19), or $42/19 = 2.21$.

As shown in Table 8.1 (page 92), it would probably be best to feed this amount of fluoride from a 4 per cent (saturated) solution of sodium fluoride by means of a solution feeder.

To compute the required rate of delivery of the feeder, we must compute the fluoride available from the sodium fluoride saturator. A 4 per cent solution contains 4 g 100 per cent pure sodium fluoride per 100 ml water. This supplies 1.8 g fluoride ion per 100 ml water ($4.0/2.21 = 1.8$). From the computation in line 2 above it was shown that 1.63 g fluoride ion per min are required to fluoridate 575 gal water per min to obtain a level of 0.75 ppm. We would then need $1.63/1.8 \times 100 = 91$ ml of 4 per cent sodium fluoride solution per min.

Another way of computing this rate would be to consider the problem as a proportion:

$$\frac{0.75 \text{ ppm}}{18{,}000 \text{ ppm}} = \frac{\text{ml of 4 per cent solution required per min } (x)}{575 \text{ gpm} \times 3{,}785 \text{ ml per gal}}$$

Then, solving for the unknown, we get

$$(x) = \frac{1{,}640{,}000}{18{,}000} = 91 \text{ ml per min}$$

It can be seen from the descriptions of the various types of solution feeders that almost all manufacturers can supply a feeder of this capacity. With the fluoride compound and the type of feeder selected, it is then necessary to choose the particular make, from the point of view of accuracy, pressure requirements, ease of adjustment, maintenance requirements, and costs. When this is determined, the various pieces of auxiliary equipment, described in Chap. 9, should be considered.

CHAPTER 9 *Feeder Auxiliary Equipment*

Feeder Control

After the type and size of the feeder have been determined, but before ordering, a decision will have to be made as to how the feeder should be controlled. Inasmuch as fluorides should be fed more accurately than any other water-treatment chemical, considerable thought should be given as to how the most accurate feeding can be economically accomplished.

Such control can be either manual or automatic. Manual control can be very accurate under ideal conditions but requires constant vigilance and a high order of skill to obtain a long-sustained, constantly accurate fluoride level. Automatic control, on the other hand, generally results in more accurate feeding. It also frees, for other duties, the personnel who would have to monitor the feeder performance. Being an inherently more accurate method, automatic control provides a safer system for feeding fluorides and tends to conserve chemicals by preventing overfeeding. The provision of automatic control, however, is somewhat expensive and requires additional maintenance.

The methods described in this chapter for automatic control encompass many different systems and principles, involving combinations of mechanics, hydraulics, electronics, and chemistry. They may sound bewildering to the layman, but in the water-treatment and chemical industries they are among the "tools of the trade."

Manual control is obtained by adjusting the feeder by hand for every change in conditions which may affect the final fluoride level. The most common and pronounced of these conditions is a change in the quantity of water to be treated. However, many other causes for a change in fluoride levels may arise. These include changes in the purity of the fluoride compound, changes in the dissolving characteristics of the chemical, and changes in the fluoride level in the raw water. In any case, the fluoride concentration must be periodically determined, and a corresponding adjustment must be made by hand in the setting of the feeder. This system is used very widely in the larger water plants, where adequate analytical laboratory services are continuously available, where expert operators are constantly in attendance for adjusting the feeders, and where few, if any, unpredictable changes in the quantity and quality of raw water occur.

Two systems of automatic control are available. Inasmuch as the quantity of water is the most commonly encountered variable and usually has the greatest effect on fluoride levels, most control systems are based on an adjustment of the feeder automatically, depending on the quantity of water to be treated. The other system of automatic control of fluoride feeders is based entirely on the fluoride level in the treated water. This, of course, requires the continuous automatic analysis of the fluorides in the treated water and a means for adjusting the feeder to maintain a constant, preset fluoride level. The apparatus required to do this is described in Chap. 11.

In the first of these two systems, commonly known as "pacing," the feeder is automatically forced to keep up with the amount of water to be treated: increasing the amount of fluorides fed when the quantity of water to be treated increases and lowering the amount fed when the water rate falls.

A pacing system requires the provision of three pieces of equipment in addition to the feeder: (1) a sensing device to determine the extent of the change in the quantity of water to be treated, (2) a transmission system for conveying a signal which indicates the necessity for changing the feeder setting, and (3) a means for changing the output of the feeder.

Primary Devices (Meters). It is a difficult problem to select the most economical primary device to measure the amount of water to be treated with the greatest accuracy commensurate with cost. Generally, there is one best device for this purpose for each installation, and much study should be devoted to selecting it. Factors which affect the type of device to be selected include the quantity of water to be measured, the accuracy desired, the quality of the water, the pressure of the water at the point of measurement, the maximum

Feeder Auxiliary Equipment

variations in flow, the size of the pipeline, the amount of pressure loss which can be afforded in passing through the measuring device, and the space available for the installation. Brief descriptions of the most popular of the devices available for measuring various quantities of water are given; many others are omitted because they are primarily used for measuring liquids containing suspensions or other impurities or for measuring oils, air, sewage, or other fluids.

Measuring devices which are at present not ordinarily used by water utilities include pitot tubes, variable area meters, "parshal" or other open flumes, and weirs. Descriptions of these are omitted not only because such devices are too inaccurate for our purposes, but also because many of them are primarily intended for measuring liquids other than water. Those that will be considered include displacement, current, and propeller meters, orifice plates, venturi tubes, Dall and Gentile tubes, nozzles, and magnetic-flow meters. Except for the displacement, current, propeller and magnetic-flow meters, the operation of all is based on the principle of the creation of a head differential between the upstream and downstream sides of the restricting measuring device.

The most commonly used device for measuring water quantities is the displacement (oscillating-piston or nutating-disk) water meter. The principle of operation is the same for both types. A measuring chamber is fitted with a piston (Fig. 9.1) or disk (Fig. 9.2) which rotates within it (in the case of the disk the motion is a combination of rotary and wobble, i.e., nutating). For each revolution of the piston a definite quantity of water is displaced. The rotating motion of the piston is converted to some convenient unit of measurement (cubic feet, cubic meters, gallons, etc.) on a dial by means of appropriate gear trains. Meters of this type are available from $\frac{1}{2}$-in. pipe size ($\frac{5}{8}$-in. body) up to 6 in. The corresponding nominal delivery rates are from 12 to 700 gpm. However, the vast majority of these meters are made in the $\frac{1}{2}$- or $\frac{3}{4}$-in. sizes for measuring the amounts of water used in homes, small apartments, and hotels. They are used primarily where the range in flow is from the very smallest to fairly high. Their outstanding characteristic is their ability to measure accurately the smallest flows. Accuracy is in the order of $\frac{1}{2}$ to $1\frac{1}{2}$ per cent. For this reason, they are ideally suited for pacing small fluoride feeders when used with the smallest, home-sized private water supplies or the smallest water-treatment plants. Their cost varies from

$30 for the smallest size up to $1,500 for the 6-in. size. The head loss for the 6-in. meter at 700 gpm is 10 psi.

Velocity, turbine, or current meters are of two types: those primarily intended for the premises of water consumers and those for use in water plants or pumping stations. They can be readily used interchangeably, depending on the quantities of water to be measured. As with displacement meters, the standard specifications for

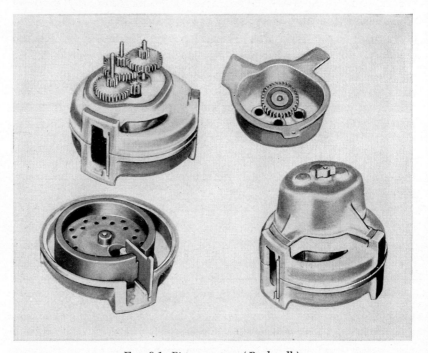

FIG. 9.1. Piston meter. (*Rockwell.*)

these meters have been formulated by the American Water Works Association. Current meters for use where smaller quantities of water are involved (Fig. 9.3) are described in AWWA Specification C701. Meters of this type are distinguished by removable measuring chambers. Also, the turbine usually rotates in a horizontal plane, and the register on top is directly connected to the turbine shaft. The impact of the flowing water on the blades of the turbine rotates the turbine shaft at a speed proportional to the flow of water. The shaft is connected to the register through a gear train, and the indicated quantity of water is proportional to the revolutions of the turbine.

Feeder Auxiliary Equipment

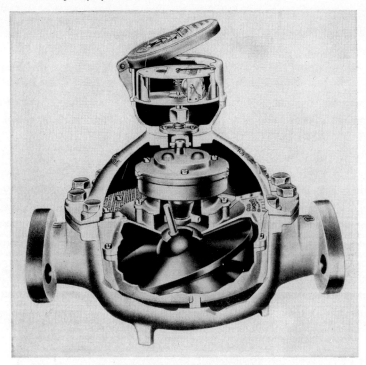

Fig. 9.2. Disk meter. (*Rockwell.*)

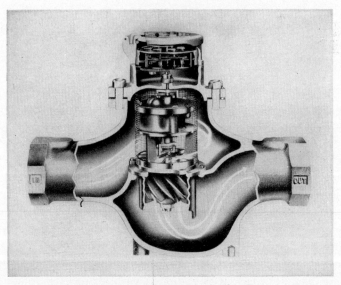

Fig. 9.3. Current meter. (*Rockwell.*)

Turbine meters are made in sizes from 1½- to 20-in. pipe size. The more generally used sizes fall between the range from 2 to 10 in., corresponding to nominal flows of 250 to 5,500 gpm. For sizes larger than 6 in., however, the propeller meters may be found to be more economical, not only in first cost, but also in lower head losses. Advantages of current meters are that they can pass large quantities of water with less damage to the working parts and with a comparatively low head loss. On the other hand, such meters are of little value for measuring the smaller flows. For instance, a 2-in. current meter cannot measure flows accurately below 10 gpm and will not register at all below 2 to 3 gpm. They are ideal for moderate, relatively constant rates of flow, such as those from an unthrottled well pump. Their accuracy is in the order of 2 per cent. As with other mechanical meters, careful maintenance is required to obtain optimum accuracy. A turbine meter with a standard register, for 4-in. pipe connections (maximum capacity 900 gpm), costs about $500. As with displacement meters, pacing is accomplished by adding an electric contactor or drive rod to the register shaft.

Turbine and displacement meters are made by:

Badger Meter Manufacturing Company, Milwaukee, Wisconsin
Buffalo Meter Company, Inc., Buffalo, New York
Hersey-Sparling Meter Company, Dedham, Massachusetts
Neptune Meter Company, New York, New York
Rockwell Manufacturing Company, Pittsburgh, Pennsylvania
Well Machinery & Supply Company, Fort Worth, Texas
Worthington Corporation, Harrison, New Jersey

Compound meters, combining a displacement and a current meter in the same casing, are used where wide variations in flow must be measured (Fig. 9.4). Inasmuch as they must be equipped with two registers (and consequently with two electric contactors), they are not ordinarily used for pacing fluoride feeders. The complexity of the electrical accessories necessary to convert the electrical contacts to periods of feeder motor operation results in a rather expensive installation. If such an installation cannot be avoided, it may be more economical to provide a separate feeder for each electric contactor.

The propeller meter, the other type of current meter, is described in AWWA Specification C704; it is generally inserted directly into the pipeline (Fig. 9.5). It is equipped with a propeller rotating in a

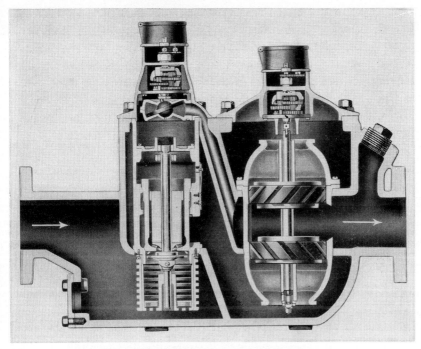

Fig. 9.4. Compound meter. (*Badger.*)

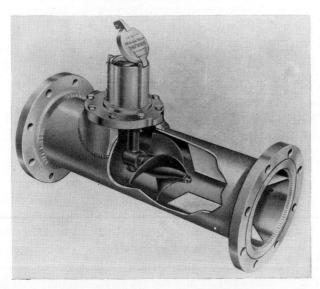

Fig. 9.5. Propeller meter. (*Sparling.*)

vertical plane and in the flowing stream of water. The speed of rotation is assumed to be proportional to the velocity of the water, and the register, driven through gears, is designed to indicate the quantity of water passing the propeller. Propeller meters are available in sizes ranging between 2 and 72 in. with capacities of 80 to over 60,000 gpm. Each propeller meter has a definite minimum recommended flow below which the accuracy deteriorates. A 6-in. meter, for instance, should not be used for flows below 90 gpm. For this reason this type should never be installed where it is desired to record or indicate flows which even occasionally are below the minimum recommended.

It is also necessary to design the piping near propeller meters so that the line containing the meter will always flow full of water. Accuracy is affected when the pipeline is flowing only partially full. With proper use and careful maintenance the error should be no greater than 2 per cent. This type of meter is particularly desirable where large quantities of water are to be measured and where relatively high accuracy is desired. It is also particularly easy to install in that the meter body is essentially a short section of pipe. It is available for any type of pipe joint (flanges, screwed, bell-and-spigot, welded). If pipe joints cannot be used, such a meter can be installed in existing lines by inserting the propeller and its shafting through a hole cut into the pipe and fastening the assembly with bolts or welds to the pipe. The cost is about $250 for the 2-in. size and $500 for the 6-in.

Like the other types of meters, these can be fitted with electric contactors or mechanical-hydraulic pacing mechanisms connected to fluoride solution or dry feeders. Details of these devices are described below. Propeller meters are made by:

B-I-F Industries, Providence, Rhode Island
Measure-Rite Meter Company, Alhambra, California
Sparling Meter Company, El Monte, California

In all the meters described above a vertical spindle is rotated to drive an indicator dial (the register) through a gear train. This spindle can also be made to drive three types of mechanisms to pace fluoride feeders. (1) A parallel, vertical shaft can be driven by a set of gears on which is mounted a small cam (Fig. 9.6). As this cam rotates, it makes and breaks an electric circuit by moving the contact arm of a microswitch or mercury switch. (2) In addition, an auxiliary vertical shaft, driven by a gear train, can be used to drive a small crank which causes a valve rod to reciprocate. This valve permits water (or compressed air or any other actuating fluid) to enter the driving chamber of a reciprocating piston or diaphragm type of solution feeder. Solution feeders are generally used in connection with these types of primary devices because of the relatively smaller quantity of water involved. (3) The motion of the vertical shaft in the

FIG. 9.6. Electric contactor fitted to disk meter. (*Hersey*.)

meter can be changed through miter gears to rotary motion of a horizontal shaft, which protrudes through the case of the meter. The end of this shaft is fitted with a cam for actuating a hydraulic valve which controls the water driving a diaphragm type of solution feeder.

The electrical signal produced by the first of these methods is utilized in many different ways to vary the output of a fluoride feeder:

1. The impulse caused by closing the microswitch is used to operate a solenoid (magnetic) valve (Fig. 9.7). The valve might be used to permit water or air under pressure to actuate either a diaphragm-driven feeder (Fig. 9.8) or an air-motor-operated valve (Fig. 9.9) on the outlet line of the dissolver of a dry feeder.

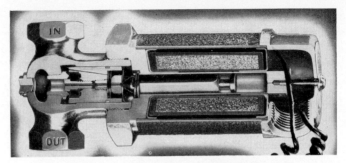

Fig. 9.7. Solenoid valve. (*AKTOmatic.*)

2. The impulse provided by the microswitch may be used to drive directly the shaft of a magnetically actuated diaphragm feeder.

3. Either through electric relays or directly (depending on the power requirements) the impulse from the microswitch may be used to actuate a magnetic clutch or a belt shifter on the shaft of a motor driving a feeder.

Fig. 9.8. Hydraulically actuated feeder. (*B-I-F Industries, Inc.*)

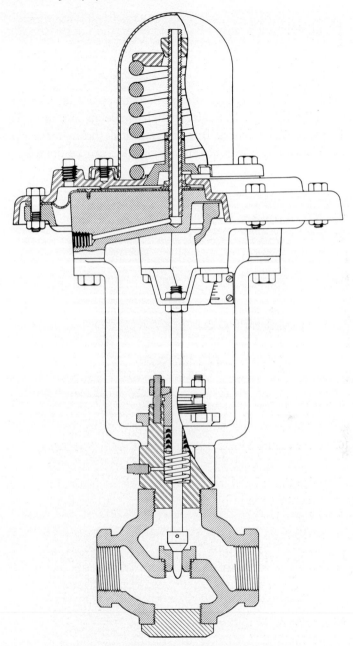

Fig. 9.9. Air-motor-operated valve. (*Foxboro.*)

4. If the impulses from the microswitch are too frequent for the starting and stopping ability of the motor on the feeder, the impulses can be accumulated by a ratchet-type relay, which permits the motor to run a preselected period of time after each group of impulses has been received.

5. If it is desired to control the periods of operation of an electric motor directly through the microswitch, some means must be provided to prevent the feeder from operating if the meter happens to stop at the "make" position of the switch. This can be done by insert-

Fig. 9.10. Feeder control by meter-actuated water valve. (*B-I-F Industries, Inc.*)

ing an impulse timer between the meter and the feeder motor. By this means each contact made by the microswitch drives the feeder motor a preselected number of seconds or minutes.

The second means for using this type of meter for pacing (operating a pneumatic or hydraulic valve) is shown in Fig. 9.10, where a hydraulically operated feeder is used. As shown, this feeder has the same fluoride solution end as the motor-driven models described in Chap. 8, but it is driven by a diaphragm from the energy derived from water or from air under pressure.

The third method (utilizing the rotating motion of a horizontal shaft from the meter head) is shown in Fig. 9.11. The rotating shaft in this case operates a cam which opens a water valve leading to the operating diaphragm of a reciprocating solution feeder. Many types of water meters can be equipped with this horizontal shaft.

Other methods for controlling a feeder from an electric impulse from a water meter are available: controlling the speed of a direct-current motor; automatically (with a reversing electric motor or by air cylinder) controlling the stroke length of the feeder; or controlling with an air cylinder or reversing motor the position of the belt (and therefore the speed of the feeder) of a variable-speed-drive motor. These means are generally more complicated and expensive than those described.

FIG. 9.11. Feeder control by shaft from meter head. (*Wallace & Tiernan.*)

Pressure-differential Primary Devices. Pressure-differential metering systems usually consist of three basic units: a pressure-differential producer; a transmission unit to provide the recorder or indicator with the signal produced by the primary device; and the recorder or indicator to measure the amount of differential pressure and translate this difference into units of quantity or rate.

The operation of these devices depends on the principles propounded by G. B. Venturi (1746–1822). The measuring device is a restriction placed in a water line which causes an increase in velocity of the water passing through it, with an accompanying drop in pressure. The magnitude of the pressure difference between the upstream and the downstream sides of the restriction is proportional to the quantity of water passing the obstruction. The pressure difference is converted by manometers or electrical or pneumatic devices to any convenient unit of measurement in the recorder or indicator. The obstruction producing the pressure difference can be an orifice plate, various types of tube sections (venturi, Dall, or Gentile tube), and nozzles (Fig. 9.14).

The most common and simplest device in waterworks practice is the orifice plate (Fig. 9.12). It can be installed in a pipeline of any size under any conditions of pressure. If it is properly installed and the line is flowing full, the upstream and downstream pressure differential will be reproducible with an accuracy within 0.5 per cent 95

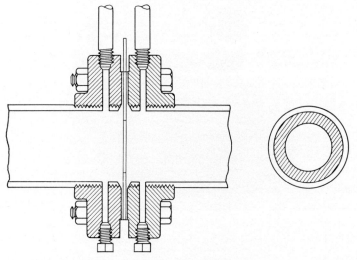

Fig. 9.12. Orifice plate. (*Foxboro.*)

per cent of the time (this is a definition of accuracy usually used with primary devices). The installation cost of orifice plates is lower than for any other type, since it is merely necessary to insert the metal plate into the line, usually between two flanges. In addition, their first cost is by far the lowest. They can be made readily by any good machinist from flat plate stock. For these reasons, orifice plates are used more often than any other primary device. However, where the highest accuracies at large flows are desired or where only the lowest loss of head can be tolerated, the tube-type differential pressure producers are preferred. Orifice plates for a pipe diameter of 6 in., made of stainless steel, cost $30. The integrator-recorder may cost $600.

They can be obtained from B-I-F Industries, Inc., Providence, Rhode Island, and Foxboro Company, Foxboro, Massachusetts.

The venturi tube, invented by Clemens Herschel in 1887, is one of the simplest and most efficient primary devices. As shown in Fig. 9.13, it is inserted into and forms a part of the pipeline. The size of the line may be any diameter between ½ in. and 10 ft. As the flow of liquid

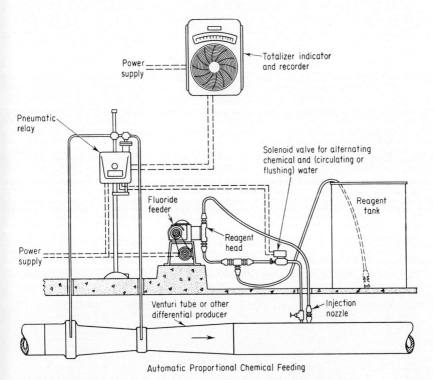

Fig. 9.13. Automatic proportional chemical feeding using venturi tube. (*B-I-F Industries, Inc.*)

travels through the inlet portion of the tube, the smaller diameter produces a rapid increase in velocity with a corresponding decrease in pressure (see Fig. 9.14). After the area of minimum diameter (the throat) is reached, the flow gradually decelerates until the full diameter of the pipe is regained. Chambers constructed at the entrance section and at the throat connect to the outside of the tube and lead to the signal-transmission system. The operation of insert nozzles, Dall tubes (see Fig. 9.15), and Gentile tubes is similar to that of venturi

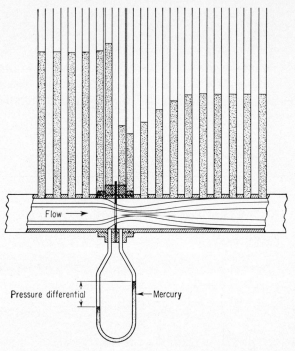

Fig. 9.14. Pressure variations through a pressure-differential primary device. (*Foxboro.*)

Fig. 9.15. Dall tube. (*B-I-F Industries, Inc.*)

Feeder Auxiliary Equipment

tubes. Venturi tubes require a longer laying length (a 12-in. venturi tube, for instance, may be as long as 8 ft) and are considerably more expensive. (A 6-in. Dall tube, for instance, costs $250.) Their use can many times be economically justified, however, by the savings realized in pumping costs because of their low head loss. A 6-in. venturi tube costs $600. The same type of indicator recorder required for orifice plates costs $600. Such tubes are manufactured by:

B-I-F Industries, Providence, Rhode Island
Foster Engineering Company, Union, New Jersey
Foxboro Company, Foxboro, Massachusetts
Infilco, Inc., Tucson, Arizona
Minneapolis-Honeywell Regulator Company, Minneapolis, Minnesota
Simplex Valve & Meter Company, Lancaster, Pennsylvania

Instruments for indicating or recording the flow through pressure-differential devices are made by:

Bailey Meter Company, Cleveland, Ohio
B-I-F Industries, Providence, Rhode Island
Fischer & Porter Company, Hatboro, Pennsylvania
Foxboro Company, Foxboro, Massachusetts
Hagan Corporation, Pittsburgh, Pennsylvania
Infilco, Inc., Tucson, Arizona
Minneapolis-Honeywell Regulator Company, Minneapolis, Minnesota
Republic Flow Meters Company, Chicago, Illinois
Simplex Valve & Meter Company, Lancaster, Pennsylvania

The magnetic-flow meter, while not used widely in waterworks practice at present, has certain inherent advantages which may warrant serious consideration for water measurements in the future. It consists of a short section of nonmagnetic pipe, electrodes on the inner surface of this pipe, an electromagnet surrounding the pipe, and a means for conducting, amplifying, and indicating the current generated. The meter is essentially an alternating-current generator wherein the flowing, conductive water induces a voltage when passing through the magnetic field. The voltage is proportional to the velocity of the water, and the recorders and indicators are readily adaptable to reading any units of volume. Their extreme accuracy (in the order of 1 per cent of full scale), together with the accuracy of the recorder, makes many installations accurate to the reading of 1.5 per cent. In addition, they are not affected by density and temperature of the water or by slurries, nor do they require a straight run of pipe upstream. A 6-in. magnetic-flow meter with recorder costs about $3,000.

Signal Transmission

Chemical feeders are almost invariably located some distance from the primary device because of the various operating requirements described in Chap. 10. In some cases this distance may be measured in miles, although generally both the sensing device and the feeders are located within the same building. The signal generated by the primary device may be sent to the feeder electrically, pneumatically, hydraulically, or mechanically. In addition, means are available whereby these systems can be changed from one to the other. For instance, if the primary device produces an electric impulse (as in a water meter equipped with an electric contactor), equipment can be obtained that will permit this signal to be changed to a pneumatic signal and transmit such a change in air pressure to the actuating device on the feeder.

The most common primary devices used in water plants are those employing a difference in water pressure. They measure flow indirectly by producing a change in the velocity head, and by measuring this difference, the velocity of the water through a constant cross-sectional area can be inferred. Knowing the cross-sectional area of the passage where the pressure difference is observed, one can compute the volume of water passing through.

This difference in pressure is transmitted by piping to a mercury manometer. Here (Fig. 9.16) the rise and fall in the mercury level produced by the pressure difference cause a float to rise and fall a proportional distance. The float in turn is connected by means of shafts and gears to a pointer or pen, which indicates or records the flow at any time. The difference in pressure can also be transmitted through piping to various types of force-balance cells (Fig. 9.17). The controlled air pressure from such devices not only is used for recording, but can be made to control the output of feeders of any type.

The distance from the primary device to the manometer should be as short as possible—less than 75 ft. For longer distances, the signal must be transmitted either pneumatically or electrically; at present the pneumatic system appears to be more reliable and economical. In addition, the maintenance of pneumatic-transmission equipment is generally simpler and the components are less likely to be affected by the humid atmosphere generally found in water plants.

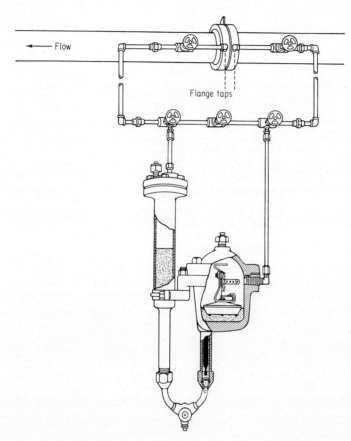

Fig. 9.16. Mercury-manometer installation. (*Foxboro.*)

Manufacturers supplying pneumatic-transmission equipment also provide electrical equipment. These include:

Askania Regulator Company, Chicago, Illinois
B-I-F Industries, Providence, Rhode Island
Bristol Company, Waterbury, Connecticut
Foxboro Company, Foxboro, Massachusetts

Minneapolis-Honeywell Regulator Company, Minneapolis, Minnesota

Simplex Valve & Meter Company, Lancaster, Pennsylvania

Pneumatic Transmission. A detailed description of the differences in these devices cannot be given here. The subject is highly technical, and full coverage of the differences would in itself fill an entire book. The devices are all basically similar, however, in that all use the same

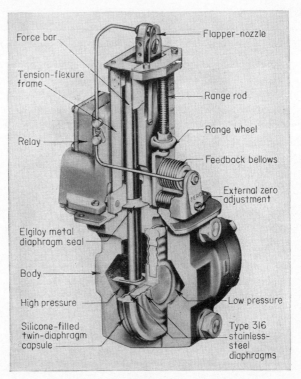

FIG. 9.17. Pneumatic relay. (*Foxboro.*)

operating principle. The air pressure in the pneumatic-transmission line between the primary device and the manometer or chemical feeder is controlled so that it is essentially proportional to the flow through the primary device. In all instruments the proportional control of the air pressure is obtained by using a double restriction in the air line, a flapper, and a pneumatic relay (Fig. 9.18).

A pressure-regulated source of clean, dry instrument air (at 20 psi pressure) is necessary for the operation of this system of control. This air is passed through orifice resistance 1 and flows to the atmosphere through orifice 2. The amount of air escaping depends on the position of the flapper—close to or away from the nozzle. The fulcrum of the flapper is connected to the pointer of the pen arm of the indicator or recorder, which is positioned by the float in the manometer. As the position of the flapper changes, more or less air is released from the air system and the pressure (increased by means of a pneumatic relay) to the receiver (a diaphragm, piston, or other air-actuated device) changes in almost direct proportion to the position of the flapper. In this manner the signal generated by variations in flow of water can be transmitted fairly long distances (in some cases 600 ft) to the actuating mechanism (diaphragm-operated valve, bellows, air piston) controlling the output of the fluoride feeder. This is accomplished by changing the position of a rheostat ahead of a motor, a

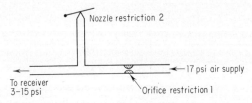

FIG. 9.18. Basic elements of pneumatic transmission. (*Foxboro and Water Pollution Control Federation.*)

speed changer (belt, gears, or disks), the length of stroke on the feeder, and many other similar means. In addition, the same principle is utilized for controlling a fluoride feeder according to the fluoride content of the treated water. This is described in Chap. 11. Similar equipment is used for transmitting, recording, and indicating many other measurable variables; viz., temperature, pressure, water or chemical level in tanks or bins, viscosity, force, position, humidity, pH, velocity, and others.

Electrical Transmission. Any task performed by or required of a pneumatic system can also be accomplished by utilizing a large variety of electrical components. The simplest form of electrical transmission and control involves the use of electric contactors on water meters. An electric impulse is produced at the end of the passage of a certain quantity of water through the meter. The frequency of contacts is governed by the selection of the gear ratios and shape of the rotating cam, which actuates the microswitch or mercury switch of the contactor. The impulses thus produced are used to drive a feeder either through a solenoid type of drive directly connected to a reagent head or through a time-delay relay which starts and stops a motor or drive shaft on a feeder.

If, instead of an electrical-contact-equipped water meter, the primary device used is a differential-pressure producer, then the pressure is so controlled as to position a bellows, diaphragm, or manometer float. The position so obtained can be converted to an electrical signal by either potentiometers, differential transformers, capacitors, induction coils, oscillators, photometers, synchro motors, cam-operated switches, or combinations of these. All have particular advantages, but in selecting the best for a particular application, a careful study should be made of advantages and prices of the applicable components. In many cases the final selection will be based on the economy resulting from utilizing components similar to existing equipment. The manufacturer of the equipment should be consulted first in this regard.

Pneumatic versus Electrical Transmission. The basic question, of course, is which system should be selected before any instruments are considered—those operated by compressed air or those controlled by electricity. Both, in fact, work almost equally well in most applications. At present pneumatic instruments predominate for short transmission distances, but the overall proportion is gradually changing in favor of electrical. Most instrument manufacturers are now making both types for similar tasks. The principal advantages of the electrical systems are faster response time and longer transmission distance. If a telemetering system, which is essentially an impulse-duration type of signal transmission system, is available (for recording and integrating flows), then its signal can be used directly for pacing a feeder. This system is commonly preferred in waterworks. Pneumatic systems are limited to about 600 ft, and the air lines are subjected to clogging with moisture, particularly in freezing temperatures. The disadvantages of electrical instruments are that they generally cost more and that they do not as yet have a standard signal. Pneumatics usually have a standard 3- to 15-psig signal, which most manufacturers use. This makes all pneumatic instruments compatible and readily adaptable in any pneumatic system. Electricity, on the other hand, comes in a wide variety of voltages (either alternating or direct current), frequencies, and other current characteristics, so that economical interchangeability is impossible. There is also considerable criticism of the quality of the final control elements; i.e., the lack of a really reliable electrically operated valve. Electrical instruments may also be subject to corrosion, short-circuiting, and other failures because of the humid atmosphere to which many of them are exposed in a water plant. There does not appear to be agreement on which system requires less maintenance. Because electrical components are in many instances mass-produced for the electronics industry, however, they are generally less expensive than parts for pneumatic systems.

Means for Changing Output of Chemical Feeders

The output of a feeder can be changed in two basic ways:

1. By changing the speed of rotation, nutation, or pulsation
2. By controlling the quantity of chemical delivered per stroke, revolution, or pulsation

The speed of rotation, nutation, or pulsation can be changed in any of the following ways:

If electrical (electric motor):

1. By changing the voltage applied to a direct-current motor driving the feeder, the speed of rotation is changed. Thyratron tubes are used to convert alternating current to direct (Fig. 9.19).

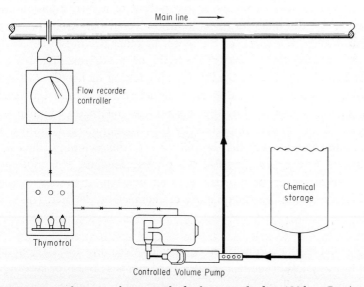

Fig. 9.19. Voltage-regulation method of pacing feeder. (*Milton Roy.*)

2. By means of engaging and disengaging a magnetic clutch driving a feeder and thereby starting and stopping the feeder during controlled intervals with the motor running continuously.

3. By means of an electrically actuated belt shifter where the driven sheave drives the feeder. Here, too, the motor runs continuously.

4. By changing the setting on a variable transmission. Such changes in speed transmission can be made either electrically by means of small motors, pneumatically, or hydraulically by means of power cylinders.

5. By starting and stopping a motor on a timed cycle or other electrical contacting device. This system (impulse-duration) is commonly a part of conventional integrating flow meters with an electric relay to actuate the feeder motor. The relay may also be connected to a three-way solenoid valve on the suction line to the feeder. When the solenoid is set for feeding, the feeder is connected to the source of fluoride solution; when set for not feeding, the feeder pumps only plain water.

If pneumatic or hydraulic control is available:

1. The frequency of stroke is controlled by varying the pressure of an actuating fluid.

2. The length of stroke is controlled by position of the piston in a power cylinder or reversible motor.

The second way of changing the output of a feeder automatically is to change the quantity of material delivered per stroke or per pulsation, leaving the speed of the motor or other driving force constant. These devices are generally electrical or air-driven. They are designed to change automatically the cam setting or length of stroke of a solution feeder, depending on a signal received from the primary device. On a dry feeder, the control can be made to change the degree of movement imparted to the trough on a vibratory type volumetric feeder or the amplitude of the scraping devices on oscillating-tray or rotating-disk volumetric feeders. This system can also control the opening of the gate on belt-type gravimetric feeders, which in turn controls the quantity of chemical on the belt.

Feeder Accessories

The auxiliary equipment necessary or desirable for the operation of fluoride solution feeders is different from that used for feeding dry compounds of fluorides. It is necessary, for instance, for solution

Feeder Auxiliary Equipment

feeders to be connected to a tank where the solutions are prepared and stored. In addition, there might be provided a means for recording the quantity of solution used during a particular time interval, and equipment must be available for the safety of the water-plant operators.

The simplest tanks for holding solutions are those used for preparing sodium fluoride solutions of about 2 per cent strength (see Chap. 7). Such a tank is usually made of stainless steel or, more recently, of various plastics and is of such size that it holds sufficient solution to last at least one day. Means should be provided for measuring the quantity of water added to the tank from a faucet installed above the tank. A mixer, either electric or a hand paddle, should be used long enough to dissolve all of the fluoride compound. A small scale is necessary to weigh the sodium fluoride to be dissolved in each measured quantity of water. A good spring scale is suitable for this purpose. A sight glass installed in the side of the tank is useful for checking the delivery of the solution feeder (Fig. 8.17).

For obtaining saturated solutions of sodium fluoride, without the necessity of measuring either the water or the fluoride chemicals, a sodium fluoride saturator is used. This device, described in Chap. 8, can be obtained ready-made, complete with float-controlled water inlet, from the B-I-F Industries, Inc. A water meter is usually installed on the water line so that the quantity of 4 per cent sodium fluoride solution withdrawn can be measured at regular intervals and the concentration of fluoride in the treated water computed from the quantity of water treated.

To prevent or reduce the formation of calcium or magnesium fluorides (which are relatively insoluble) in the saturator, feeder, solution lines, and injection nozzle, it is sometimes desirable to pretreat the water used in the saturator or in a dry-feeder dissolving chamber. The source of this calcium or magnesium is from the dissolved compounds in the water used to make the fluoride solution, and in order to prevent formation of this scale or precipitate, the calcium or magnesium can be removed or combined. This can be done either by softening the water (see Fig. 8.1) or treating it with one of the hexametaphosphates (Calgon, Micromet, or Nalco 918). From the viewpoint of the economic loss of fluorides due to the formation of calcium or magnesium fluorides in a saturator, a softener is not advisable for water whose hardness is less than about 80 ppm. In many cases, clogging of a saturator for this reason can be prevented by periodically backwashing the precipitate out of the tank. When injector nozzles (the fittings used at the end of the discharge hose of a solution feeder and inserted through a corporation cock into the pipeline at the point of fluoride application) become clogged with this scale, they can be withdrawn and cleaned periodically.

A tank of special design (described in Chap. 7) for dissolving fluorspar (CaF_2) in alum solutions of various concentrations is used in plants where coagulation is required. Plants of almost any size can use this system if the minimum coagulant dosage is in the order of 10 ppm when the amount of fluoride ion to be added is 1.0 ppm.

When hydrofluosilicic acid is used, rubber-lined steel tanks are necessary, not only for bulk storage of the acid, but as reservoirs and weighing (day) tanks for supplying the acid solution feeder. Tanks made of ceramic materials are unsuitable because of the formation of hydrofluoric acid at the surface of the liquid. This acid dissolves the lining of such tanks. The hydrofluosilicic acid is transferred to the measuring tanks by means of pumps or by gravity. Pumps are fitted either with Hastelloy or Carpenter 20 steel. Plastic pipelines have been used successfully in many places.

The quantity of acid used is readily determined by recording periodically (daily or hourly) the weight of acid fed. This is determined by mounting the shipping container or acid weigh tank on platform scales. Such readings are necessary in order to compute the actual average fluoride concentration, which is based on the amount of water treated and the amount of acid used between successive, periodic weighings of the acid. The same computation should, of course, be used with all fluoridation systems, whether liquid or dry fluoride compounds are used.

The beam of the platform scales can be connected to a weight recorder which will trace the amount of fluoride compound used throughout any period. Such records are invaluable for establishing the continuity and accuracy of fluoride feeding.

For safely handling fluoride compounds (discussed in more detail in Chap. 12), the water-plant operators should be supplied with equipment designed to protect them from the toxic effects of inhaling or otherwise ingesting such materials. For handling powdered and dusty dry chemicals rubber gloves, masks, and aprons should be available. The masks may be of the type approved for toxic dusts (BM2101) and fitted with filters for such masks (BM2133). They are available from the Mine Safety Appliances Company, Pittsburgh, Pennsylvania. For hydrofluosilicic acid the operators should wear rubber gloves, rubber aprons, and boots and should have ample water for dilution in case of spills.

Dry-feeder installations, being larger, are more complicated and expensive than similar equipment for feeding liquids. Dry-feeding systems usually consist of a combination of two or more devices consisting of material conveyors, hoppers, dust collectors, scales and recorders, and means for transferring the fluoride solution from the dissolving chamber of the feeder to the point of application.

Conveyors, which are found only in the largest plants, are used for moving the fluoride compounds from the railway or truck unloading platforms to the storage hoppers. They are designed for either mechanical or pneumatic operation. The mechanical conveyors are a combination of bucket elevators and screw or belt conveyors. They are usually totally enclosed in a tight metal housing to prevent spreading dust. Such conveyors are made by:

B-I-F Industries, Providence, Rhode Island
Bucket Elevator Company, Summit, New Jersey
Hapman Conveyors, Inc., Kalamazoo, Michigan
Jeffrey Manufacturing Company, Columbus, Ohio
Link-Belt Company, New York, New York

Pneumatic systems involve the movement of chemicals in a pipe by means of air under either pressure or vacuum. The chemicals are removed with a suction hose from the railway car or truck and blown into receiving tanks. The spent air is passed through a "cyclone" separator or filter to remove the dust prior to its discharge to atmosphere. Pneumatic systems are made to order by:

Allen-Sherman-Hoff Company, Wynnewood, Pennsylvania
Daffin Manufacturing Company, Lancaster, Pennsylvania
Day Company, Minneapolis, Minnesota
Dracco Corporation, Cleveland, Ohio
Fluidizer Company, Hopkins, Minnesota
Fuller Company, Catasauqua, Pennsylvania
Spencer Turbine Company, Hartford, Connecticut

Hoppers for holding dry chemicals, made of wood, concrete, or metal, are generally either built into the structure of the water plant or made to order to fit the space between the dry feeder and the floor above. Depending on their size, configuration, and type of chemicals stored, they may be equipped with dust-control equipment and vibrators (to move the compound to the outlet) or bag loaders (Fig.

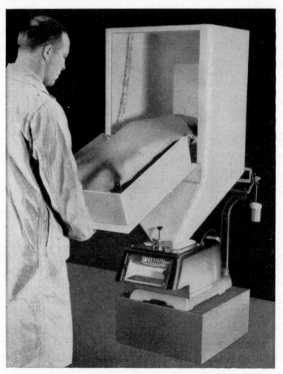

Fig. 9.20. Bag loader. (*B-I-F Industries, Inc.*)

9.20). Some manufacturers combine bag or barrel loading equipment with dust-control devices. These can be obtained from:

B-I-F Industries, Providence, Rhode Island
Ducon Company, Mineola, New York
Metals Disintegrating Company, Summit, New Jersey
Pangborn Corporation, Hagerstown, Maryland

Vibrators for the hoppers are available from the Eriez Manufacturing Company, Erie, Pennsylvania, and the Syntron Company, Homer City, Pennsylvania.

Feeder Auxiliary Equipment

Most dry feeders can be equipped by the manufacturers with clock-actuated charts to record the quantity of material being delivered from the feeder. Such a device should be considered for all installations, not only for obtaining a record of the amount fed but for calculating (knowing the amount of water treated from meter readings) the concentration of fluoride maintained in the treated water.

After the fluoride chemical has been dissolved in the water in the dissolving chamber (see Chap. 8) of the dry feeder, it must be removed and added to the main stream of water at the point of application. This is done either by gravity (Fig. 9.21), with an eductor, or

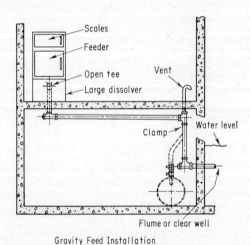

Fig. 9.21. Gravity flow of fluoride solution. (*B-I-F Industries, Inc.*)

with a pump. If done by gravity into an open channel, no additional equipment except for the piping is required.

Usually an eductor or pump is required to raise the fluoride solution to a level higher than the feeder or to introduce it into a line under pressure. For their operation, eductors require water under a pressure of three to four times the discharge head. In order to prevent the introduction of air into the main line, both the eductor and the pump should draw the fluoride solution from a suction well which is supplied continuously with water through a float-controlled valve. The strength of the fluoride solution diluted in this manner will have no effect, of course, on the fluoride concentration in the treated water.

Other accessories which are available and may in some cases be desirable include bell or light alarms which indicate over- or underfeeding; flow indicators or controllers on the water line to the dissolving chamber to assure the proper ratio of water to fluoride com-

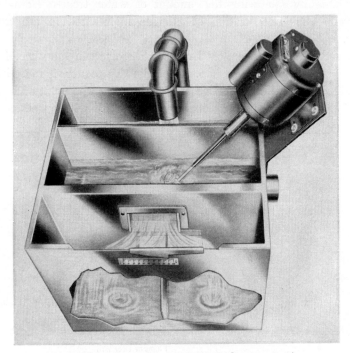

FIG. 9.22. Flow splitter. (*B-I-F Industries, Inc.*)

pound; carts, trucks (motor- or hand-driven), pallets, and pallet loaders for moving and storing bagged fluoride compounds; weir tanks and flow splitters (Fig. 9.22) for dividing accurately the effluent from the dissolving chamber where more than one point of fluoride application is necessary.

CHAPTER 10 *Points of Application*

Fluorides are almost always in the form of a liquid just prior to being added to a water supply. These solutions of fluorides may be obtained from the dissolver of a dry feeder, from a saturator or other type of dissolving tank in which sodium fluoride or other soluble fluorides are used, or in the form of hydrofluosilicic acid. The relative difficulty of selecting a point at which fluoride solutions will be applied to the water depends on whether the water is otherwise treated (coagulated) or untreated (except for chlorination).

If the supply is untreated, such as that obtained from wells, the fluoride solution is generally fed with a solution feeder or a pump (with a suction tank) directly into the discharge line of the well pump. Fluoride solutions are never fed directly into the well or into the suction side of the well pump without some provision made to prevent siphoning. When a fluoride solution feeder is designed to discharge at a lower level or into the suction side of a pump, an air gap or a float-controlled suction box (Fig. 10.1) should be provided. Without such a device to prevent siphoning of fluoride solution through the solution feeder, gross inaccuracies in the fluoride level in the treated water are possible.

Other factors being equal, the preferred location should be chosen from the point of view of absence of excessive pressures, convenience of replenishing the fluoride compounds, and ease in measuring the quantity of water to be treated. Many communities are supplied by numerous wells all discharging at different points directly into the distribution system. In most cases, because it is economically prohibitive to join the discharge lines of most of the wells, each well must be considered a separate fluoridation installation and designed as if it serves, by itself, a separate community.

In systems using water otherwise treated, the preferred point of application is sometimes a compromise between the place where the feeder hopper can be replenished easily and the point where subsequent treatment processes will produce the least tendency to remove the added fluorides. These processes include lime softening in the presence of magnesium; alum or sodium aluminate coagulation; feeding of bentonitic clays; and activated-carbon treatment at low pH values. As much as one-third of the applied fluorides may be removed from the treated water when the alum dose is 100 ppm. It is therefore generally advisable to add fluorides after these treatment steps. (If

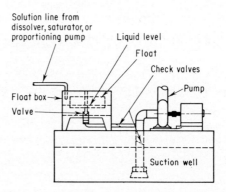

FIG. 10.1. Float-controlled suction well. (*B-I-F Industries, Inc.*)

settling is prolonged and efficient and if the filters remove relatively little material, fluorides can be added before filtration.) If the load on the filters is considerable, however, fluorides should be added after filtration, generally either in the line between the filters and the clear well or into the clear well itself.

In some few instances, it is economically justifiable to realize some loss of fluoride. Where transportation, storage, or handling costs of the fluoride compounds can be substantially reduced by adding fluorides to the raw or partially treated water at a more accessible point, the loss of fluorides through the plant may be small enough, in some cases, to result in a substantial saving. For instance, at Washington, D.C., fluorides are added to the raw water for the McMillan plant. This is more economical (despite a loss of about 0.1 ppm fluoride through subsequent treatment) than loading trucks with the fluoride

Points of Application

compound, driving them to the plant, unloading them, and providing space and facilities for storage and feeding equipment.

Besides the advantages of locating the feeder near the best place to store the chemicals, the advantages of a short fluoride solution line (between the feeder and the point of application) are also important. Longer lines increase pumping costs and require considerable additional effort when the accumulated sludge is periodically removed.

It is also a definite advantage to be able to permit the liquid fluorides from the dissolving chamber of a dry chemical feeder to flow by gravity into the clear well or open channel (Fig. 9.21). This generally requires that the feeder be located above these structures and rather near. In any case, the point to be preferred is where the pressure to be overcome (by the solution feeder) is the least.

Chlorine, chlorine dioxide, or chloramines can be added anywhere in the plant or distribution system. There are no known reactions between fluorides at 1.0 ppm in water and the compounds formed in water after chlorination. The only untoward effect of chlorine and its compounds is the bleaching effect on the reagents used in the fluoride determination, which makes the samples read too high (samples appear to contain more fluoride than they actually have). This problem and the means for overcoming it are discussed in Chap. 11.

Even though a well supply is not chlorinated, the addition of a fluoride solution would not ordinarily provide a source of bacterial contamination of the supply. Sodium fluoride solutions, for instance, are germicidal at only 5,000 ppm (0.5 per cent as sodium fluoride), and most such solutions are prepared considerably stronger.

CHAPTER 11 *Control of Fluoride Concentration (Laboratory Procedures)*

Planning for any fluoridation installation should include provisions for periodic testing of the treated water for fluoride concentrations. A complete record of the history of the fluoride levels made not only by the local community water and health departments but by the state health department laboratory should be maintained. Such a record is useful for any future questions of liability for damages alleged to result from over- or underfeeding of fluorides, for correcting the settings on the fluoride feeding equipment, and for proving the performance of the waterworks in contributing to the health and welfare of the community.

Determining periodically the fluoride content of water samples is one of the two ways available to prove the adequacy of fluoridation. The result of such laboratory testing for fluorides reveals the concentration at the instant the sample was collected. A more accurate and revealing method of collecting a sample would be to utilize equipment which would provide a composite sample of the entire day's output of treated water. Similar results, useful in furnishing a check on the laboratory results, are obtained by computing the fluoride dose for each day (or longer periods of time). This computation is based on the quantity of fluoride compound used and the amount of water treated. For instance, if a fluoride-free water is being fluoridated with 98 per cent pure sodium fluoride at the rate of 1,000,000 gal per day and 18.0 lb sodium fluoride was used during the same period, then the fluoride concentration in ppm would be:

$18.0 \times 98\% = 17.5$ lb pure sodium fluoride used

$\dfrac{17.5}{2.21} = 7.9$ lb fluoride ion (2.21 is the molecular ratio of NaF to F ion; see Chap. 7)

1 ppm = 8.3 lb fluoride ion per million gal water

Then, average dose for this day would be $\dfrac{7.9}{8.3} = 0.95$ ppm

A sample of water taken at any time during the day should, when analyzed, contain within 0.1 ppm of the fluoride concentration found by this computation.

The quantity of fluoride compound used in the computation is obtained from readings of the scale on which the dry feeder or chemical hopper rests, from the recording of the poise position on the beam of gravimetric feeders, from the measurement of the level in fluoride solution tanks, from the meter readings on the line replenishing sodium fluoride saturators (assuming an invariable production of a 4 per cent solution), and from the loss of weight of the hydrofluosilicic acid shipping containers or day tanks. The quantity of water treated is obtained from reading the master meter, which should be standard equipment in every water plant.

FREQUENCY OF SAMPLING

The number of samples taken for fluoride analysis is chosen with the primary objective of assuring a constant fluoride level at the correct concentration throughout the distribution system. It is on the basis of such determinations that the fluoride feeder is adjusted and its accuracy judged.

The number of samples collected depends primarily on one or more of these factors:

1. The size of the system: the larger the water plant, the more samples required.

2. The complexity of the system: if, for instance, more than one fluoride feeder is necessary, then additional samples will have to be collected.

3. The age of the installation: more samples are required during the initial stages of a fluoridation project primarily in order to check the settings on the fluoride feeder.

Although it is obviously impossible to specify the exact number of samples required for all plants, the following general observations can be made:

1. For the smallest communities, at least one sample should be examined each day.

2. During the early stages of a project (the first six months or so) at least one sample at the water plant and one additional sample from a remote point in the distribution system should be examined.

3. In larger communities the health department laboratory usually runs check samples along with the water department.

4. In larger communities a few (up to a dozen or more) distribution samples are taken each day from a series of perhaps a hundred or more sampling points. The places at which the daily samples are taken are rotated or changed each day among the total points available.

5. The records of all sampling should be preserved and stored so as to be readily available.

FLUORIDE ANALYSIS

The analysis of fluoride in water requires the determination of the quantity of fluoride ion present in solution regardless of the source of that ion. It may come from many sources; it may occur naturally from the fluoride compounds in the earth or from the fluoride or silicofluoride compounds added in controlled amounts at the water plant. There is no method for distinguishing one fluoride ion from another or for determining its source or the exact nature of the original fluoride compound (see Chap. 1).

Of the large number of analytical methods for fluorides, the method chosen must be of such sensitivity as to detect very minute quantities of fluoride present in a sample. The usually desired fluoride level (1.0 ppm) involves only about $\frac{1}{10}$ oz in 1,000 gal water. For this reason most of the traditional chemical methods of analysis (gravimetric, volumetric, polarographic, etc.) are unsuitable. The so-called colorimetric methods are the most suitable because they are sensitive to fluorides at the low levels encountered in water and are relatively easily performed with the simplest types of laboratory glassware and instruments.

The methods of analysis of all the usually occurring elements in water, including fluorides, which are considered the most practicable

Control of Fluoride Concentration (Laboratory Procedures) 161

and accurate are described in the book "Standard Methods for the Examination of Water and Wastewater," published jointly by the American Water Works Association, Federation of Sewage and Industrial Wastes Associations, and American Public Health Association; the cost of this book is $10. Whenever possible the methods contained in this book should be used. The standard fluoride methods were developed on the principles of colorimetric analysis.

Colorimetric analysis is based on the relationship of the color intensity of a solution to the concentration of a particular substance in the solution. By an appropriate system of measurement this relationship can be used as a quantitative analytical method. The colorimetric system utilizes three basic components: a source of radiant energy (light); a sample-positioning device; and a receptor or detector of

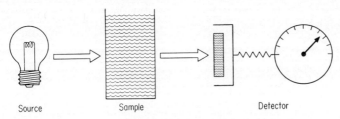

Fig. 11.1. Color-measurement system. (*USPHS.*)

radiant-energy values (Fig. 11.1). All colorimetric systems, regardless of their complexity, depend essentially on the manipulation of these three parts.

In all systems of colorimetry, light of a particular quality is made to pass through a colored sample. The degree of coloration of the sample is proportional to the concentration of a particular ingredient —in this case, fluoride. The intensity of the light which is not absorbed on passing through the sample (the transmitted light) is measured by an appropriate detector, and it is then compared to the intensity of transmitted light previously measured on samples containing known concentrations of the ingredient sought (fluoride).

In fluoride analysis the colorimetric system employs white light as the radiant-energy source. The simpler systems use sunlight through a convenient window or a ceiling or desk light indoors. The more sensitive but complicated photometers use a tungsten filament bulb with a regulated voltage source of current to ensure a constant light intensity.

The positioning of the sample must be the same in a series of light-intensity measurements. The samples are placed in either Nessler tubes (for visual comparison) or carefully prepared ground-glass or quartz cells or cuvettes when comparisons are made in a photometer. The cells must be of such a design that the length of the light path is identical for each measurement.

The detector is used to measure (and eventually to indicate) the intensity of light passing through the sample. In the simplest systems the detector is the human eye; the more accurate instruments utilize photoelectric cells. The human eye is generally considered less accurate because many people are color-blind to varying degrees and the accurate comparison of two colors depends greatly on judgment and experience.

In almost all color-measurement systems, a reagent is selected which, when added to the water sample, produces a color, the intensity of which is indicative of the concentration of the element sought.

COLOR-MEASUREMENT SYSTEMS

Visual methods of color comparison are the oldest, simplest, and most widely used. Nessler tubes are almost invariably employed as the sample container in this method. In fluoride analysis the so-called long-form tubes, designed for a 30-cm light path, should be obtained. The tubes should be bought in sets of six or more and should be matched with each other from the point of view of length of light path and color of glass. In this system outside daylight or room light is the radiant-energy source, the tube itself is the sample container, and the eye is the detector.

As shown in Fig. 11.2, the tubes are held so that light is passed lengthwise through the sample and standards. Two or more tubes are held side by side, and the sample is matched as closely as possible with the standards which contain known concentrations of fluoride.

The principal advantages of the Nessler tubes are that they are inexpensive and very simple to use. A matched set of six tubes costs $8.50, and a stand or rack to hold them (or to facilitate making color comparisons) costs $6 (Fig. 11.3). More elaborate racks with built-in light source are also available. They are easy to read, inasmuch as the long light path makes color comparisons more precise. In the fluoride determination, the chemical reaction which causes the color development can be carried out within a tube which will be used later for color comparison.

The disadvantages in Nessler tubes are derived principally from the vagaries of the human eye. These include various degrees of color

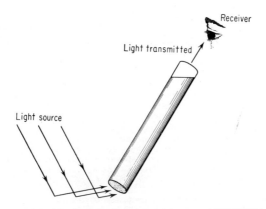

FIG. 11.2. Nessler-tube measuring system. (*USPHS.*)

blindness, degree of fatigue experienced by the analyst, and his training and experience. Accuracy is lost also when the comparisons are made in the extreme ends of the spectrum (in the red or purple areas). In addition, the standards and samples must be prepared and compared simultaneously because the eye cannot "remember" and later compare shades and intensities of color.

Also available are devices for visual color comparison which contain permanent color standards either in the form of solids (colored glass) or of liquids confined in sealed glass vials. The advantages of these instruments are that standards do not have to be prepared for each batch of samples and a light source may be included as standard equipment. The latter feature reduces some source of error when variable natural light must be used. Permanent standards are avail-

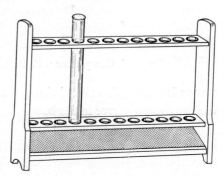

FIG. 11.3. Nessler tube in use with rack. (*Fisher Scientific Co.*)

able for a large number of chemical determinations, particularly in the water, sewage, and industrial-waste analytical field.

A slide comparator containing liquid standards in vials is shown in Fig. 11.4. Here the center sample tube contains the fluoride reagent, and the resulting color is compared with the color of the two fluoride concentrations on either side. The two standard colors are viewed through tubes containing distilled water or water having a turbidity similar to the sample. Light from either a natural or an artificial

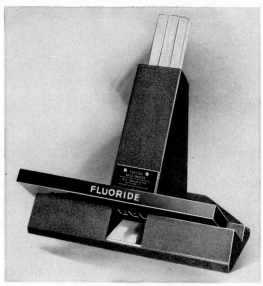

FIG. 11.4. Comparator with liquid color standards. (*Taylor.*)

source is passed downward through the three tubes and is reflected to the eye from a mirror in the base.

A disk type of color-standard comparator is shown in Fig. 11.5. Here the varied colored-glass standards are fitted in a disk which, when rotated, permits comparison of various fluoride concentrations with the sample, one standard at a time. An eyepiece containing a double prism permits the two colors to be viewed in juxtaposition. Disks for many other determinations are available for this instrument.

The disadvantage of permanent color-standard instruments is the inherent risk of error resulting from the faults of the eye as a color comparator. In addition the deterioration or misuse of the reagent may make color matching extremely difficult. If the manufacturers' directions are strictly followed (particularly those relating to reagent measurement, reaction time, and temperature), an accuracy of about 0.2 ppm at the 1.0-ppm fluoride level can be expected. These instruments can therefore be used successfully to check the error which might result from malfunctioning of the chemical feeder, but even then the instruments should be checked periodically with samples containing known concentrations of fluoride. The outstanding advantage of this group of instruments is that standards do not have to be prepared for each group of samples. Such instruments are made by the W. A. Taylor & Company, Baltimore, Maryland, and Hellige, Inc., Garden City, New York. They cost between $35 and $100, the latter including a built-in light source.

Fig. 11.5. Comparator with glass color standards. (*Hellige.*)

PHOTOMETERS

The use of photometers for comparing samples with standards eliminates or substantially reduces the errors contributed by the human eye in visual color-comparison methods. These instruments indicate, by electrical and mechanical means, the amount of light transmitted (or absorbed) by a colored solution. The detector in these instruments is a device (Fig. 11.1) wherein a change in electric current occurs as the light impinging on it changes.

Photometers are generally more sensitive and accurate than the human eye because the quality (wavelength) of the light, prior to passing through the sample, can be accurately controlled and then, on passing through, can be more accurately measured. For instance, the curves shown in Fig. 11.6, which apply to the Megregian-Maier reagent, represent the amount of light transmitted by solutions containing 0.0 and 3.0 ppm fluoride when using light of various wavelengths. If readings were made with a color comparator using light essentially in the 450-mμ band, no distinguishable difference in transmittancy could be obtained because the difference in the amount of light transmitted is very small. However, if a wavelength were chosen where this difference in transmittancy was greatest (i.e., at 525 mμ), the maximum sensitivity would be realized with a resulting improvement in accuracy.

A photometer consists of five essential parts: (1) a light source, usually a tungsten-filament bulb with some means for maintaining a

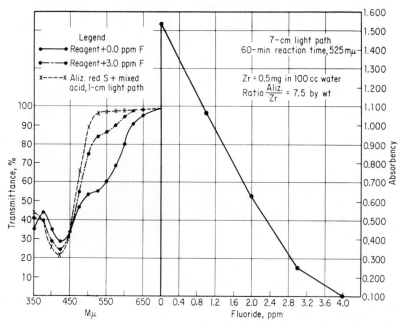

FIG. 11.6. Light transmission versus fluoride concentration. (*USPHS.*)

constant intensity; (2) a means for selecting a band of wavelengths either with filters (Fig. 11.7) or, as is done in spectrophotometers (Fig. 11.8), with prisms or diffraction gratings; (3) an optical device for concentrating, splitting, or directing the light beams; (4) means for holding and exposing to the light beam the sample and standards; and (5) a system for sensing the transmitted light and indicating its intensity. This combination of parts makes it possible for a photometer to measure, with the greatest accuracy, the amount of light passing through a colored solution.

Photometers range in price from $150 to over $5,000. Generally, the higher-priced ones are more accurate and complex and are intended for research or other purposes which require extreme precision. The lower-priced instruments have in most instances been found quite suitable for fluoride analysis and for almost all other determinations required in water-plant laboratories.

Usually filter photometers (Fig. 11.7) are cheaper than those equipped with prisms and diffraction gratings. Even though filter photometers do not ordinarily produce narrow wavelength bands (except when equipped with expensive interference filters) in fluoride

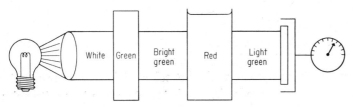

Fig. 11.7. Filter photometer. (*USPHS*.)

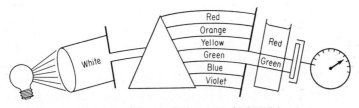

Fig. 11.8. Spectrophotometer. (*USPHS*.)

and other water analytical methods, such narrow bands are not necessary. For instance, on the curves shown in Fig. 11.6, almost equal sensitivity can be obtained between wavelengths of 500 and 550 mμ. It would then ordinarily be unnecessary to obtain a very much more expensive instrument in order for a waterworks laboratory to obtain a bandpass of, say, 525 mμ, plus or minus 5 mμ, from which could be obtained only a slight increase in sensitivity.

In addition, it is sometimes advantageous to use so-called "split-beam" photometers if local conditions produce voltage variations in the electric-current source. These variations cause a change in the intensity of light in the bulb, which, of course, changes the amount of light transmitted. Any variation in the intensity of the light source is automatically compensated by the split beam; i.e., the variations in light intensity are the same whether the light passes through the sample or to the other receptor.

The type of receptor (barrier-layer cells or phototubes) seems to have little bearing on the accuracy of fluoride analysis. Generally phototubes are used on the more expensive instruments, but the tubes must be supplied with an outside source of current and their output is usually amplified. However, barrier-layer cells have been found to be just as well suited for fluoride analysis and appear to be just as sensitive as phototubes.

The operation of all photometers is essentially the same. The directions of the manufacturer should be followed when balancing the galvanometer and in adjusting it for obtaining readings in the area of the galvanometer scale, which produces the widest reading range per unit of fluoride. As with any other colorimetric method, the color of the sample is compared with the color of known concentrations in similarly prepared standards. In photometers, the transmittancies of a series of standards are read and plotted on graph paper. Curves are obtained for each batch of fluoride reagent used and also whenever there is reason to suspect a change in the transmittancy readings of the standards. A method for plotting such results for fluoride is shown in Fig. 11.6: the concentration of fluoride is usually plotted along the horizontal axis; the absorbancy or transmittancy on the vertical axis. In order to obtain a straight line, the readings should be converted and plotted as the logarithm of the transmittancy. Some instruments indicate this logarithmic value on the galvanometer scale directly. With these data, the transmittancy values of subsequently prepared samples can be entered on the curve and the fluoride values obtained directly. In many cases, the vast majority of fluoride readings will be made within a narrow fluoride-concentration range; i.e., between 0.9 ppm and 1.1 ppm where the optimum is 1.0 ppm. In such instances, it may save time and possibly improve accuracy to compute the fluoride concentration directly on a slide rule without resorting to a standard curve. For instance, if the transmittancies of the 0.9- and 1.1-ppm fluoride concentrations are 190 and 170 on a certain photometer and the sample reads 175, then the fluoride in the sample would be: 0.9 plus 0.15 = 1.05 [twenty transmittancy units are equivalent to 0.2 ppm fluoride; then 15 units (190 − 175) are equivalent to 0.15 ppm].

Control of Fluoride Concentration (Laboratory Procedures)   169

Photometers are available from the following sources:

Filter Photometers

Bausch & Lomb Optical Company, Rochester, New York
Coleman Instruments, Inc., Maywood, Illinois
Epic, Inc., New York, New York
Fisher Scientific Company, Pittsburgh, Pennsylvania
Hach Company, Ames, Iowa
Hellige, Inc., Garden City, New York
Instrument Development Laboratories, Attleboro, Massachusetts
Klett Manufacturing Company, New York, New York
E. Leitz, Inc., New York, New York
Photovolt Corporation, New York, New York
Rubicon Company, Philadelphia, Pennsylvania

Spectrophotometers

Bausch & Lomb Optical Company, Rochester, New York
Beckman Instruments, Inc., South Pasadena, California
Coleman Instruments, Inc., Maywood, Illinois
Photovolt Corporation, New York, New York

CHEMISTRY OF FLUORIDE ANALYSIS

The basis of a successful colorimetric analytical method is the formation of a color which will reveal the amount of the desired constituent. This requires the availability of a reagent which will form a color when added to a sample. The intensity of this color indicates the quantity of the material sought—in this case, fluoride.

With the exception of the so-called SPADNS reagent all the fluoride methods in "Standard Methods" are based on the zirconium-alizarin reaction (Fig. 11.9). The reaction between the compounds containing the zirconium and alizarin produces a red-colored lake, or complex. Fluoride in the sample removes some of the zirconium from the reaction, preventing some of the red lake from forming and consequently decreasing the intensity of the color. In samples particularly high in fluorides, the color is the same as the unreacted alizarin; in low-fluoride samples, the color is similar to the red zirconium-alizarin lake. The reaction is not instantaneous but progresses with time. At the end of an hour or so the reaction is almost completed. For this reason, the samples are read 60 ± 2 min after the reagents are added. Inasmuch as the temperature of the samples affects the reaction rate, the standards and samples should be held at the same temperature during the hour.

The Scott-Sanchis reagent, used for visual comparison, contains both sulfuric acid and hydrochloric acid; this makes it particularly tolerant to higher levels of chlorides and sulfates in the sample. The

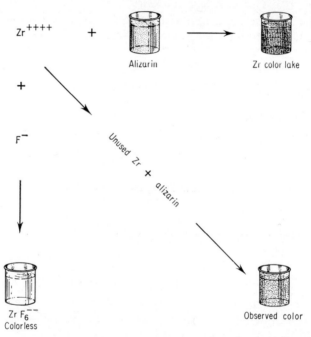

FIG. 11.9. Zirconium-alizarin reaction in fluoride analysis. (*USPHS*.)

Megregian-Maier reagent, containing more alizarin and zirconium, produces deeper colors and makes their evaluation possible on photometers. The SPADNS reagent contains, besides zirconium, a different dye called sodium 2-(p-sulfophenylazo)-1,8-dihydroxy-3,6-naphthalene disulfonate and a higher acid concentration. The SPADNS method is particularly desirable because the reaction is instantaneous, and consequently the samples can be read immediately after the reagents are added. As shown in Table 11.1, the SPADNS reagent is not so tolerant to interference (particularly sulfates)

Control of Fluoride Concentration (Laboratory Procedures)

TABLE 11.1. INTERFERENCES
CONCENTRATION OF SUBSTANCE, IN PPM, REQUIRED TO CAUSE
AN ERROR OF (+) OR (−) 0.1 PPM AT 1.0 PPM

	Scott-Sanchis	Megregian-Maier	SPADNS
Alkalinity	400(−)	325(−)	5,000(−)
Al^{+++}	0.25(−)	0.2(−)	0.1(−)*
Cl^-	2,000(−)	1,800(−)	7,000(+)
Fe^{+++}	2(+)	5(−)	10(−)
$(NaPO_3)_6$	1.0(+)	1.1(+)	1.0(+)
PO_4^{---}	5(+)	5(+)	16(+)
SO_4^{--}	300(+)	400(+)	200(+)
Chlorine	Must be completely removed with arsenite		
Color and turbidity	Must be removed or compensated for		

* This figure is for immediate reading. Allowed to stand two hours. Tolerance is 3.0 ppm. Four-hour tolerance is 30 ppm.

as are some of the other fluoride reagents, although it is particularly resistant to the effects of chlorides and alkalinity.

FLUORIDE ANALYTICAL PROCEDURES

I. Scott-Sanchis colorimetric method
   A. Reagents
      1. Acid-zirconium reagent
         a. Dissolve 0.3 g zirconyl chloride ($ZrOCl_2 \cdot 8H_2O$) or 0.25 g zirconyl nitrate $ZrO(NO_3)_2 \cdot 2H_2O$ in 50 ml distilled water in a 1-liter flask.
         b. Dissolve 0.07 g sodium alizarin monosulfonate (certified dye) in 50 ml distilled water.
         c. Add the alizarin solution to the zirconium solution in the flask, while swirling. Allow the resulting solution to stand for a few minutes.
         d. Dilute 112 ml concentrated HCl to 500 ml with distilled water.
         e. Carefully add 37 ml concentrated H_2SO_4 to 400 ml distilled water and dilute to 500 ml. After cooling, mix the two acids.
         f. To the clear zirconium-alizarin solution in the 1-liter flask, add the mixed acids to the 1-liter mark and mix. Allow to stand 1 hr before use. When stored in the refrigerator, the reagent is stable for at least two months.

2. Fluoride stock solution
   a. Dissolve 0.2210 g NaF in 1 liter distilled water. (1 ml = 0.1 mg F)
3. Standard fluoride solution
   a. Dilute 100 ml of the above stock solution to 1 liter. (1 ml = 0.01 mg F)
4. Sodium arsenite solution
   a. Dissolve 1.8 g $NaAsO_2$ in 1 liter distilled water.

B. Procedure
1. Prepare a series of fluoride standards by pipetting the indicated amounts of the standard fluoride solution into 100-ml Nessler tubes and diluting to mark.

Ml standard solution	Mg F	Ppm F
0	0.00	0.0
2	0.02	0.2
4	0.04	0.4
6	0.06	0.6
8	0.08	0.8
10	0.10	1.0
12	0.12	1.2

2. Take 100 ml of the sample, or an aliquot diluted to 100 ml, in a 100-ml Nessler tube and dechlorinate by adding 2 drops (0.1 ml) of the arsenite solution for each ppm chlorine present in the sample.
3. Adjust the standards and sample to the same temperature, within ±2°C.
4. Add 5 ml of the mixed reagent (above) to each of the standards and sample. Mix well and allow to stand at least 60 min.
5. Compare the sample with the standards and determine the amount of fluoride present in the sample aliquot.
6. Report fluoride concentration in ppm.

II. Megregian-Maier photometric method
A. Reagents
1. Alizarin: 0.740 g sodium alizarin monosulfonate (certified dye) dissolved in 1 liter distilled water.
2. Zirconium solution
   a. Dissolve 0.354 g zirconyl chloride ($ZrOCl_2 \cdot 8H_2O$) or 0.294 g zirconyl nitrate ($ZrO(NO_3)_2 \cdot 2H_2O$) in about 100 ml distilled water.
   b. Add 100 ml concentrated HCl.

c. Dilute to about 600 ml.
d. Add 33.0 ml concentrated H_2SO_4 slowly with stirring.
e. Cool, then dilute to 1 liter with distilled water. Let stand 1 hr (or over night) before use.
3. Fluoride stock solution: 0.2210 g NaF dissolved in 1 liter distilled water. (1 ml = 0.1 mg F)
4. Standard fluoride solution: dilute above stock solution 1:10 with distilled water. (1 ml = 0.01 mg F)
5. Sodium arsenite solution: 1.8 $NaAsO_2$ dissolved in 1 liter distilled water.

B. Procedure
1. Preparation of standard curve
   a. Prepare a series of standards by pipetting the indicated amounts of standard fluoride solution into labeled 250-ml Erlenmeyer flasks and diluting to 100 ml with distilled water.

Mg F	Ml standard solution
0.00	0
0.05	5
0.10	10
0.15	15
0.20	20

   b. Adjust all of the standards to the same temperature, within $\pm 2°C$. Record the average temperature. All subsequent determinations must be made under the same conditions.
   c. Add exactly 5.0 ml alizarin reagent to each standard. Mix immediately. At 2-min intervals, beginning with the blank, add exactly 5.0 ml zirconium reagent to each standard. Mix immediately and allow to stand 60 ± 2 min, timing the reaction from the addition of the zirconium reagent.
   d. Place a portion of each standard in a cuvette.
   e. Using only distilled water in one of the cuvettes, set the photometer at zero absorbancy (100 per cent transmittancy), using a wavelength of 525 mμ (or a filter in range of 520 to 550 mμ).
   f. Make absorbancy readings for each standard within the 60 ± 2 min time limit.
      Note: Absorbancy values when plotted against fluoride content will produce a line of negative slope. The

line should be practically straight between 0.00 and 0.20 mg. The standard curve thus plotted will serve for the subsequent determinations of fluoride concentrations, provided the reaction conditions (temperature, time, reagents) are duplicated. A new curve must be prepared whenever a new batch of either the alizarin or the zirconium reagent is prepared, or whenever a different standard temperature is used.

  2. Analysis of water sample
   a. To a 100-ml sample of water (or an aliquot diluted to 100 ml) add 2 drops (0.1 ml) sodium arsenite solution for each mg per liter of chlorine present, then add 2 additional drops.
   b. Adjust the temperature of the sample to that of the standard curve.
   c. Add exactly 5.0 ml alizarin reagent. Mix immediately, then add exactly 5.0 ml zirconium reagent. Mix and allow to stand 60 ± 2 min.
   d. Transfer a portion of the reaction mixture to a cuvette and obtain absorbancy reading, using the distilled water reference as in step e, preparation of standard curve.
   e. From the standard curve, determine the amount of fluoride present in the sample aliquot. Report fluoride concentration in ppm.

III. SPADNS method
  A. General discussion
    1. *Principle.* The reaction rate between fluoride and zirconium ions is influenced greatly by the acidity of the reaction mixture. By increasing the proportion of acid in the reagent, the reaction can be made practically instantaneous. Under such conditions, however, the effect of various ions differs from that in the conventional alizarin methods. The selection of dye for this rapid fluoride method is governed largely by the resulting tolerance to these ions.
    2. *Interference.* The interference effects of ions normally present in water are listed in Table 11.1. It should be noted that the effects vary with the fluoride concentration, sometimes being of greater magnitude at lower fluoride levels. Whenever the error contributed by any one interfering substance approaches 0.1 ppm or analysis is unknown, the sample should be distilled.

B. Apparatus
   1. *Colorimetric equipment.* One of the following is required:
      a. Spectrophotometer, for use at 570 mμ, providing a light path of at least 1 cm.
      b. Filter photometer, providing a light path of at least 1 cm and equipped with a yellow-green filter having maximum transmittance at 550 to 580 mμ.
C. Reagents
   1. *Standard sodium fluoride solution.* Prepare as for other fluoride methods.
   2. *SPADNS solution.* Dissolve 0.958 g SPADNS [sodium 2-(p-sulfophenylazo)-1,8-dihydroxy-3,6-naphthalene disulfonate][1] in distilled water and dilute to 500 ml.
   3. *Zirconyl acid solution.* Dissolve 0.133 g $ZrOCl_2 \cdot 8H_2O$ in about 25 ml distilled water. Add 350 ml concentrated HCl and dilute to 500 ml with distilled water. Equal volumes of SPADNS solution and zirconyl acid solution may be mixed to produce a single reagent.
   4. *Reference solution.* Add 10 ml SPADNS solution to 100 ml distilled water. Dilute 7 ml concentrated HCl to 10 ml and add to this. The resulting solution is used for setting the reference point (zero) of the spectrophotometer or colorimeter. It is stable and may be reused indefinitely.
   5. *Sodium arsenite solution.* Approximately 0.1 N.
D. Procedure
   1. *Preparation of standard curve.* Prepare fluoride standards in the range of 0.00 to 1.40 ppm by diluting appropriate quantities of the standard fluoride solution to 50 ml with distilled water. Add 5.00 ml each of the SPADNS solution and zirconyl acid solution, or 10.00 ml of the mixed reagent, to each standard and mix well. Set the photometer to zero absorbance with the reference solution in the cuvette, and obtain the absorbance readings of the standards immediately. Plot a curve of the fluoride-absorbance relationships. A new standard curve must be prepared whenever a fresh batch of either reagent is prepared, or whenever a different standard temperature is used.
   2. *Pretreatment of sample.* If the sample contains Cl_2, remove it by adding 1 drop (0.05 ml) arsenite solution for each

[1] Eastman Organic Chemicals #7309.

0.1 mg Cl_2 and mix. (Arsenite concentrations of 1,300 ppm produce an error at 1.0 ppm fluoride.)
3. *Analysis of sample.* Use a 50-ml sample or aliquot diluted to 50 ml. Adjust the temperature of the sample to that of the standard curve. Add 5.00 ml each of the SPADNS solution and zirconyl acid solution, or 10.00 ml of the mixed reagent, mix, and read the absorbance immediately or at any subsequent time, first setting the reference point of the photometer as above. If the absorbance falls beyond the range of the standard curve, the procedure must be repeated, using a smaller sample aliquot.

E. Calculation
1. Read ppm F from the calibration curve.

F. Precision and accuracy
1. A precision of 0.05 ppm F can be obtained if the procedure is followed carefully. Temperature of sample, interferences, and measurement of reagent affect the accuracy appreciably.

SOURCES OF ERROR

Fluoride analysis requires the greatest possible accuracy. The quantities of fluoride ordinarily found in water are so minute that an error ordinarily tolerated in other methods would produce considerable errors in percentage in fluoride analysis. The most common errors can be avoided by taking the following precautions:

1. The glassware must be clean and perfectly matched as to color. This includes all glass that will be used for comparing colors—Nessler tubes and photometer cells (cuvettes). A single fingerprint on the outside of a cuvette has been known to lower considerably the fluoride level of the sample within.

2. The reagent must be very accurately measured. The fluoride reagent contains both the ingredients for forming the colored lake and also the acid for controlling the optimum pH for color development. Pipetting techniques must be thoroughly mastered.

3. The methods all have a definite upper limit for detecting fluoride accurately. When the fluoride level in a sample approaches this limit, the sample should be appropriately diluted to the range well within this limit.

4. Color and turbidity can be compensated for in all instruments,

but the safest way to handle these is to distill the sample to remove them.

5. The time required for the reaction and the temperature at which it occurs should be observed and carefully controlled.

6. Interference by other constituents in the water is probably the most common source of error. No fluoride method is specific for fluoride; all of them are capable of measuring other ions under optimum conditions. Different interferences can be handled in various ways, but the most reliable is to distill the sample.

CONTROL OF INTERFERING IONS

Whenever the analysis of a water sample is not known, or when the quantity of interfering ions is sufficient to produce an error of 0.1 ppm fluoride or more, such interferences must be removed or their effects diminished by one of the following means:

1. The interferences may be caused by chemicals used in the water plant for coagulation (aluminum salts) or disinfection (chlorine). Such interference can be avoided by selecting a sampling point within the water plant where these chemicals have not as yet been added to the water.

2. When the fluoride level is sufficiently high, interferences can be reduced or eliminated by diluting the sample with distilled water. This reduces the interference of other ions but increases the risk of multiplying the reading error by the dilution factor.

3. If the identity and quantity of the interfering ions are known, it is possible in some instances to select the fluoride reagent which will tolerate such an interference.

4. The equivalent quantities of interfering ions may be added to the fluoride standards, thereby nullifying the effects of any changes caused by such interferences.

5. When none of these methods are applicable, or in any cases of doubt, the sample should be distilled.

DISTILLATION

Two methods of distillation are described in "Standard Methods." Both are based on the technique described by Willard and Winter.[2]

[2] H. H. Willard and O. B. Winter, Volumetric Method for the Determination of Fluorine, *Ind. Eng. Chem., Anal. Ed.,* **5:** 7 (1933).

In this process, the sample is boiled in a sulfuric acid mixture which converts the fluoride to the volatile hydrofluosilicic acid—the silica in this compound being obtained from glass beads placed in the flask or from the flask itself.

One method (Fig. 11.10) involves two flasks: one, containing distilled water, is a steam generator; the other, containing the sample and acid, receives the steam from the first flask and releases the volatile hydrofluosilicic acid. This acid, in the form of a vapor, is condensed and collected in a volumetric flask.

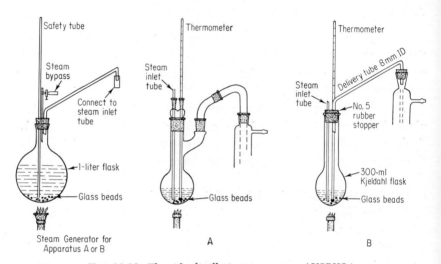

FIG. 11.10. Fluoride-distillation apparatus. (*USPHS.*)

The critical features of the apparatus are the diameter of the delivery tube connecting the flask and the condenser, the depth of immersion of both the thermometer and the steam-inlet tube, and the closeness of fit of all stoppers and joints. The delivery tube should be at least 8 mm inside diameter. A smaller tube will cause a bubble to form within it and tend to cause droplets of sulfuric acid to be carried into the condenser. The thermometer and steam-inlet tube in the distillation flask must be low enough so that both will be always immersed. When the level of liquid in the flask drops, exposure of the thermometer bulb will cause errors in temperature readings while the exposure of the steam tube will cause the steam to bypass the liquid. A leak in the joints or stoppers will cause the fluoride-containing vapor to be lost, resulting in error in fluoride readings of the distillate.

The distillation process is carried out as follows:

The 250- to 300-ml distillation flask is prepared by placing in it 125 ml distilled water, a few glass beads, and 25 ml sulfuric acid. The acid must be added slowly and mixed thoroughly with the water. After the flasks are connected as shown in Fig. 11.10, the contents of both flasks are brought to the boiling point while the steam bypass is open. The bypass is closed as soon as the temperature reaches 135°C. The burners are then adjusted so that the temperature remains between 135 and 145°C. A total of 200 ml distillate is collected at the rate of 5 to 10 ml per min. This distillate is discarded because it contains traces of fluoride which may have been in the acid or adhering to the glassware. The apparatus is now ready to receive the sample. If the sample contains very low concentrations of fluoride (less than about 0.3 ppm), it should be concentrated by boiling 200 ml down to about 50 ml. The sample should first be made alkaline with a diluted solution of sodium hydroxide if evaporation is practiced. After the distillation flask has cooled, add the sample (100 ml if not previously evaporated) to the acid remaining from the flushing-out process and mix the two thoroughly. If the sample was concentrated, distilled water may be used to rinse it into the flask. The total volume of water, however, should not exceed 100 ml. The sample is now distilled as before with exactly 200 ml of distillate collected. The acid may be used repeatedly, provided the contaminants from the water samples do not accumulate to such a degree that recovery is affected or interferences appear in the distillate. This can be detected by distilling, periodically, standard fluoride samples.

Inasmuch as the distillation process is subject to many errors, the following precautions should always be taken:

> Superheating the vapor in the distillation flask causes a carry-over of sulfates which may seriously interfere with the subsequent fluoride determination. For this reason, the flame under this flask should be kept low enough so that it does not touch the sides of the flask above the liquid level. The stoppers and joints must be tight to prevent loss of fluoride through vapor leaks. Silver sulfate must be added to the distillation flask if the chloride concentration of the sample exceeds 100 ppm. For every mg of chloride, 5 mg of silver sulfate should be added. The silver salt inhibits the volatilization of hydrochloric acid

which could interfere with the subsequent fluoride determination when using some reagents. Fluoride recovery is complete only if at least 100 ml of distillate are collected at the optimum temperature (between 135 and 145°C.). The rate of distillation should be maintained between 5 and 10 ml per minute. A rate lower than 5 ml per minute prevents complete fluoride recovery. The temperature must never exceed 145°C. because at temperatures above this point sulfate carry-over becomes serious and causes a definite interference with the fluoride determination.[3]

DIRECT-DISTILLATION PROCEDURE

The basic difference and advantage in this procedure (compared with the previous direct-steam-distillation method) is that there is no dilution of the sample. Whenever the sample must be diluted and the fluoride content subsequently determined, any error made in this determination is multiplied by the factor of the dilution. The absence of dilution is accomplished by using a larger sample and by carrying out the distillation over a broader temperature range. Because of the elevated temperature at the end of the distillation, a pronounced sulfate carry-over is experienced. However, the large volume of distillate and the lower starting temperature produce an overall sulfate concentration which is tolerated by most fluoride reagents.

The apparatus is shown in Fig. 11.11. It consists of a 1-liter round-bottom boiling flask, connecting tube, efficient condenser, thermometer (reading to 200°C) and its adapter, burner, and collecting volumetric flask. Similar apparatus may be used as long as complete fluoride recovery is obtained and the minimum sulfate carry-over is experienced. Sulfate carry-over can be reduced by the use of an asbestos shield to protect the upper part of the distilling flask from the burner flame.

In order to steam out the apparatus, 400 ml distilled water is placed in the distilling flask, 200 ml sulfuric acid is added slowly, and the two mixed thoroughly. After glass beads are added, the apparatus is assembled as shown and the joints are made tight. Heat is applied and distillation is continued until the temperature reaches exactly 180°C. This process not only removes fluoride contamination from the apparatus but serves to obtain the correct acid-water ratio for the distillation of samples. For all subsequent distillations, 300 ml of sample

[3] S. Megregian and I. Solet, Critical Factors in Fluoride Distillation Technique, *J. Am. Water Works Assoc.*, **45**(10): 1110 (1953).

Control of Fluoride Concentration (Laboratory Procedures)

are added to the previously used acid mixture, the two thoroughly stirred, and the distillation completed as described. The precautions outlined for the steam-distillation procedure apply also for this method.[4]

A recent improvement in the distillation procedure makes the process almost entirely automatic. A thermoregulator is added to the flask

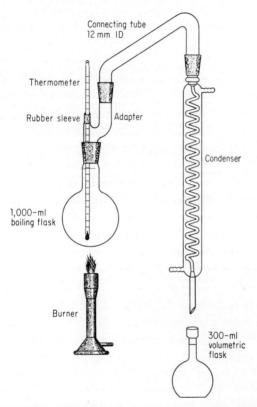

Fig. 11.11. Direct-distillation apparatus for fluoride. (*USPHS*.)

so that the heat is shut off automatically when the temperature of the mixture reaches 180°C. As a result, no attendance is necessary during distillation and the danger of exceeding the maximum temperature is minimized.[5]

---

[4] E. Bellack, Simplified Fluoride Distillation Method, *J. Am. Water Works Assoc.*, **50**(4): 530 (1958).

[5] E. Bellack, Automatic Fluoride Distillation, *J. Am. Water Works Assoc.*, **53**(1): 98 (1961).

AUTOMATIC RECORDING AND CONTROL

In Chap. 9 two systems for obtaining automatic control of the fluoride-feeding mechanisms were described. One, involving feeder adjustments based only on the quantity of water to be treated, was identified as "pacing." The other system, ordinarily known as "control," is based on the continuous analysis of the material or chemical element after it has been added to the water. The analyzed result is continuously compared to a preset value. If a deviation is detected, a correction signal is generated; this adjusts the feeder and the amount of chemical added. This loop is known as a "feedback" or "closed" system. In the present instance, this requires the continuous analysis for fluorides and a means for adjusting the feeder automatically in order to maintain a constant, preselected fluoride concentration.

Adjusting the feeder is generally accomplished by means of a recorder, which, through either electrically or hydraulically actuated signals, controls either a valve on the outlet from the dissolving chamber of a feeder, the speed or period of operation of an electric motor, the stroke length of a feeder, or many other devices or systems. For this reason, a recorder is almost always used between the sensing mechanism and the controller.

A continuous record of the fluoride concentration in water is of considerable value to a water utility. Not only would such a record monitor the operation of the entire fluoride-feeding system, but it would also provide a graphic history of the fluoride levels at any point in the plant or distribution system. It would only be necessary to provide a sensing mechanism at any desired point, and the results could readily be transmitted electrically to any monitoring or collecting center.

There are at present two promising methods for sensing automatically and continuously the fluoride level in water. One is an extension of the standard colorimetric method now in use, and the other is based on the change in the electrical conductivity of water when fluoride is added to it.

The essential parts of a continuous colorimetric method consist of (1) a sample preparation and metering system, (2) a reagent metering system, (3) a mixing and retention apparatus, and (4) photometer and recorder for indicating the fluoride level. Such an apparatus is shown in Fig. 11.12. In this particular system, the sample is not distilled and is continuously metered by an automatic pipette, the stopcock of which is turned by a timer-controlled electric motor. Similar pipettes meter the reagents. Other colorimetric instruments proportion the sample and reagents (which must be done as accurately as possible) with either capillary tubing or small laboratory chemical feeders (Fig. 11.13). The sample and reagents are then mixed and a

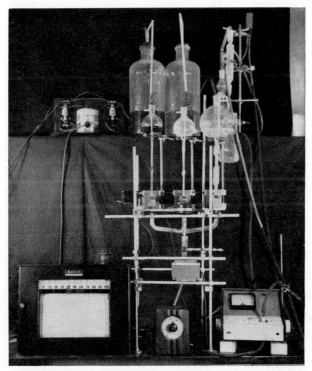

FIG. 11.12. Colorimetric fluoride-recording equipment. (*USPHS.*)

reaction time is provided in the mixing chamber. From here the mixture flows to the photometer, which is connected electrically to the recorder. Any potentiometric recorder can be wired into the galvanometer circuit of the photometer through a potentiometer. When the recorder chart is calibrated for fluoride concentration, a continuous record of the fluoride ion is obtained.

At the present time the following firms are able to supply apparatus which will continuously record fluoride levels in water with colorimetric-type instruments:

Hach Company, Ames, Iowa
Harold Kruger Instruments, San Gabriel, California
Milton Roy Company, Philadelphia, Pennsylvania
Technicon Controls, Inc., Chauncey, New York

If the interferences of the sample exceed the tolerance of the reagent selected, the sample can be continuously distilled with apparatus similar to that shown in Fig. 11.14. Only the first distillation

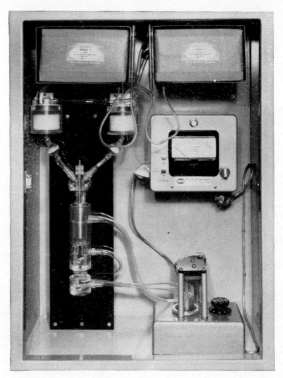

FIG. 11.13. Automatic fluoride recorder using capillary tubing. (*Hach.*)

assembly is necessary, however, because all known reagents for fluoride can tolerate the very low level of interferences found in the first distillate.

CONDUCTIVITY INSTRUMENTS

Two distillations of the sample are necessary for continuous conductivity measurements. As shown in Fig. 11.14, the second distillate is discharged directly to the conductivity cell. The second distillation is necessary because, while distillates from the first still are sufficiently free of interferences to permit direct chemical analysis, they are far from satisfactory for determinations based on conductivity. Traces of sulfates (from the sulfuric acid) and ions resulting from

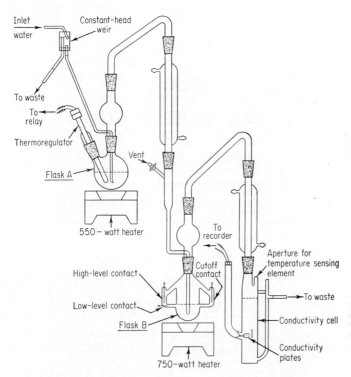

FIG. 11.14. Continuous distillation apparatus. (*USPHS.*)

volatile materials change the conductivity to such a degree that fluorides cannot be accurately measured. Consequently, the second still contains nothing but the distillate from the first still. The leads from the conductivity-cell electrodes are connected to a conductivity recorder. In this recorder the measuring circuit is a modified Wheatstone bridge with two similar air capacitors and two resistors. The sensing electrode on the conductivity cell is one of the resistors, and the balancing element in the recorder is one of the capacitors. When the sensing elements detect a change in conductivity, the measuring circuit becomes unbalanced. This sets up a voltage which, when amplified, moves the recording pen until the measuring circuit is again balanced.

Another adaptation of the conductivity principle for this purpose has been described for the Salem-Beverly water plant.[6] As shown in Fig. 11.15, two electrodes are placed in a pipeline or channel carrying water which is to be fluoridated. Between these two points the fluoride solution is added. The difference in conductivity resulting only from the presence of the fluoride compound is recorded automatically

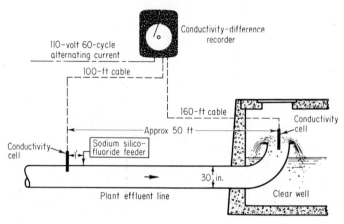

FIG. 11.15. Fluoride conductivity controller and recorder. (*Foxboro*.)

in an instrument similar to that described above. This system measures only the amount of fluoride added. No correction can be made resulting from differences in the fluoride content of the untreated water, by variation in flow of the water, or by changes in the purity of the fluoride compound.

CONTROL

Pneumatic control from any recorder is obtained by utilizing the principle described in Chap. 9. The controlled air can be used to regulate the opening in an air-motor-actuated valve. As shown in Fig. 11.16, this valve can be installed on the effluent line from the dissolv-

[6] Kenneth F. Knowlton, Investigation of the Continuous Recording of Fluoride Concentration in Water, *J. New Engl. Water Works Assoc.*, 68(1): 16 (1954).

ing chamber of a dry chemical feeder. As this valve approaches the closed position (as it would do when the fluoride level is reached or exceeded), the solution level in the dissolving chamber rises until both control electrodes are submerged. This shuts off the feeder

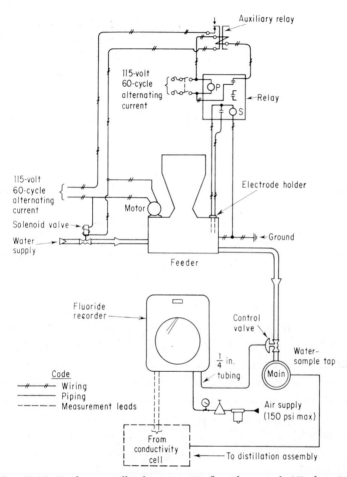

FIG. 11.16. Feeder controller for automatic fluoride control. (*Foxboro.*)

motor and also the water to the dissolving chamber. As the solution drains from the chamber, the lower electrode becomes exposed; this restarts the motor and opens the water valve. This cycle is repeated automatically and continuously.

CHAPTER 12   *Safety of the Water-plant Operators*

Waterworks operators who handle fluoride compounds must be trained to protect themselves from hazardous, accidental exposures. The toxic nature of these compounds is such that any possibility of excessive inhalation or ingestion must be prevented. The proper training of operators, particularly emphasizing the toxicity of fluoride compounds, cannot be overstressed.

Dangerous exposures to fluorides can be experienced by accidental dumping of the material (when an excessive amount might be swallowed), by inhalation from dust or vapor in the atmosphere, or by exposures which result in acid burns on the skin or eyes. No recorded exposures of such magnitudes have been reported, but the possibilities should never be minimized.

The standard for fluoride dust or vapor in air is 2.5 mg hydrogen fluoride per cu m air for a continuous 8-hr daily exposure. This level has never been even remotely approached in any water plant where concentrations in air have been measured. Fortunately, the possibilities of excessive exposure in water-treatment plants are very unlikely. Fluoride feeding equipment is generally designed so that hoppers require refilling no oftener than once a day. Inasmuch as maximum exposure is more likely to occur during such filling operations, at least it will be limited to only a few minutes a day.

Nevertheless full advantage should be taken of the procedures and equipment available to protect workmen from possible hazardous exposures. Considerable margins of safety can be obtained by choosing less hazardous fluoride compounds, by equipping feeders and hoppers with safety appurtenances, and by providing personal safety equipment.

Of the powdered fluoride compounds, fluorspar is by far the safest from a toxicological point of view. The solubility of calcium fluoride is so low that the amount absorbed by the digestive system is considerably less than from a more soluble compound such as, for instance, sodium fluoride. Safety with dry powdered chemicals can also be increased by selecting a grade or mesh size which will produce no dust and which therefore cannot be inhaled. The crystalline grade of sodium fluoride, for instance, contains very few dust-forming particles and should be used in preference to the powdered grades whenever possible. Liquid fluorides (hydrofluosilicic acid) cannot produce dust.

However, if dust-free materials cannot be used (for instance, sodium silicofluoride has an enormous economic advantage over some of the other fluoride compounds but cannot be obtained in the larger dust-free mesh sizes), then some provision must be made to suppress the dust hazard. In smaller plants, little danger to the operators is involved if reasonable care is used in handling the compounds so that the least amount of dust is produced. The greatest possibilities of exposure occur in the larger plants, where barrels of the fluoride compounds are dumped into hoppers, the openings of which are at or near the floor level. When the fluorides arch in these barrels and then suddenly empty, a cloud of dust may be formed. In all such cases the hoppers should be equipped with dust exhaust systems.

As shown in Fig. 12.1, a typical dust-collecting system forms a part of the feeder hopper. By means of a motor-driven turbine, a partial vacuum is maintained on the opening where the fluoride compound is dumped. The dust formed is drawn upward through a series of metallic or cloth screens. The filtered air is exhausted to the outside of the building. Periodically the screens are shaken or the air flow is reversed to remove the accumulated fluorides. Such devices can be obtained to handle installations of any size and are made by:

B-I-F Industries, Providence, Rhode Island
Ducon Company, Mineola, New York
Johnson-March Corporation, Philadelphia, Pennsylvania
Metals Disintegrating Company, Summit, New Jersey
Pangborn Corporation, Hagerstown, Maryland

In a few plants, the vacuum required to remove the dust from the working area is obtained from an aspirator using water under pressure. The dust in the air is entrained in the water, which is returned to some treatment point within the water plant.

A particularly desirable dust-free system for larger plants should include mechanical or pneumatic conveying systems. Such systems are particularly advantageous when carload or truckload shipments are received. The material can be transferred in a dust-free manner

Fig. 12.1. Dust-collecting equipment. (*B-I-F Industries, Inc.*)

from the shipping car to the feeder hopper without being exposed to the outside air and consequently with a minimum exposure to the operators. Such equipment is made to order and can be obtained from:

B-I-F Industries, Providence, Rhode Island
Bucket Elevator Company, Summit, New Jersey
Day Company, Minneapolis, Minnesota
Dracco Corporation, Cleveland, Ohio
Dravo Corporation, Pittsburgh, Pennsylvania

Hapman-Dutton Company, Kalamazoo, Michigan
Jeffrey Manufacturing Company, Columbus, Ohio
Link-Belt Company, New York, New York

In some of the smaller plants, a bag loader (Fig. 9.20) is used on dry feeders. With this device it is possible to dump automatically the 100-lb bags into the hopper when the door holding the bag is closed. The empty bag is removed when the next filling takes place. A dust-removal or control system is not necessary with this device.

In the very smallest plants using dry chemicals, the hopper may be replenished by using a hand scoop or shovel. Even here little hazard is involved if the operation is completed with care so that no dust is formed.

In all cases the operators should have available and wear gloves and approved masks. The following manufacturers make such masks, which have been approved by the Bureau of Mines (with the approval number shown) for protection against toxic dusts:

American Optical Company, Southbridge, Massachusetts (Approval No. 2144)
B. F. McDonald Company, Los Angeles, California (Approval No. 2126)
Hygeia Filter Company, New York, New York (Approval No. 2145)
Mine Safety Appliances Company, Pittsburgh, Pennsylvania (Approval Nos. 2107 and 2126)
Pulmosan Safety Equipment Corporation, Brooklyn, New York (Approval No. 2110)
Wilson Products, Inc., Reading, Pennsylvania (Approval Nos. 2123, 2125, and 2143)

One of the most dangerous fluoride compounds, hydrogen fluoride (hydrofluoric acid), was used for several years in an American water plant. After several accidents it was replaced by hydrofluosilicic acid. It is doubtful if any water plant in the future will find a justification for using hydrofluoric acid, and its use for this purpose should be discouraged.

Hydrofluosilicic acid is much less hazardous but should nevertheless be treated with respect. Fortunately, it is seldom necessary to handle it in any way but to pump it through pipes or tubing. If at all possible it should never be diluted before feeding (see Chap. 7). Generally, this acid can be fed directly from a shipping container or pumped directly from a storage tank. The storage tank is filled from tank trucks or railway tank cars.

The greatest care should be taken when repairing pipelines and pumps which contain this acid, particularly in protecting the hands and eyes. Goggles, face masks, rubber gloves, and boots should be worn. A copious amount of flowing water should be immediately available for flushing the eyes and for rinsing the hands and other parts of the skin.

The rooms in which acid equipment is housed should be continuously ventilated and the spent air exhausted to the outdoors. All vents on tanks should terminate outdoors in a gooseneck fitting.

In handling any fluoride compound, scrupulous cleanliness should be continuously maintained. All spills should be immediately hosed down and mopped. Acid spills can be neutralized with lime water. Thorough hand washing should be routinely practiced after each loading, and daily showers should be taken if the loadings are prolonged.

The disposal of fluoride shipping containers (barrels, cardboard drums, and paper bags) has always been a serious problem. A considerable hazard is involved, the seriousness depending on the subsequent use of the empty containers. If they are sold or given away, all traces of fluoride compound should be removed. This can be done most easily by thoroughly rinsing them with water. If they are burned on the waterworks property, the fumes should be controlled so that no nuisances are created. Fluoride-containing fumes can destroy vegetation. The safest criterion to observe is that no fluoride compound remains in any shipping container prior to its removal from the premises of the water plant.

SAFETY OF THE CONSUMER

The excellent design, construction, and ease of maintenance of chemical feeding equipment almost invariably prevents serious accidental overfeeding of fluoride compounds. There are remote possibilities, however, of overfeeding, which can be avoided by proper design of the installation.

In small solution-feeder installations, little danger of overfeeding is possible when the point of application is under pressure or at a point higher in elevation than the feeder. When injecting into a lower point or into a line under vacuum conditions, however, an open connection should be provided on the discharge line of the feeder. This condition was shown in Fig. 10.1. The suction box on the line to the suction side of the pump is necessary to prevent air from entering the pump. In some installations, when a solution feeder injects against pressure, there may be a possibility of a vacuum occurring occasionally. This may happen, for instance, when a main is being repaired in a certain part of the distribution system. If the occurrence of such a vacuum is at all possible, then the feeder should be equipped with a vacuum breaker. Many feeder manufacturers are able to supply these devices routinely with their equipment.

Vacuum breakers should also be installed on all water lines leading to dissolving chambers, saturator tanks, solution tanks, or any similar vessels holding fluoride solutions and when such a water line can be submerged under any circumstances. The safest design for such tanks terminates the water inlet above the overflow level of the tank.

Flooding of chemicals through dry feeders can be readily prevented by using a star wheel between the hopper and the dissolving chamber, as described in Chap. 8.

An obvious possibility of overfeeding could occur by permitting the fluoride feeder to operate after the water supply to be treated is shut off. A design allowing such a possibility should not be tolerated. There should always be an interconnection between the pump supplying the water and the motor driving the feeder.

A few state health departments still require that only a blue-tinted dry fluoride compound may be used in water-treatment plants, in order to lessen the possibilities of confusion with other water-treatment chemicals by the waterworks operators. The requirement, while it may have been of some value during the early periods of water fluoridation, has not proved to be entirely justifiable. The same results are attained in most states by careful waterworks management, by isolating the fluoride feeding equipment and supplies, and by better training of water-treatment-plant operators.

Periodic variations in fluoride concentration in a distribution system should ordinarily not exceed plus or minus 0.1 ppm. Such variations may occur because of pulsations imparted by the reciprocating motion in many solution feeders. This condition has been experienced when a relatively strong fluoride solution is used (hydrofluosilicic acid, for instance); when few, if any, chances of mixing the solution and the treated water occur after injection; and when some consumers are very close to the water plant. In order to level out the peaks in fluoride concentrations caused by one or more of these conditions, the following steps might be tried:

1. Provide a mixing or storage tank or basin prior to delivery of the water; or, better,

2. Increase the stroking frequency of the feeder (sometimes an additional feeder operated by the same motor is required). This, of course, will require a shorter stroking length.

3. Reduce the strength of the fluoride solution. This, of course, will require a larger volume of solution per unit of time, with an increase in stroking frequency or longer stroke lengths.

In any case, the prime consideration should always be a constant, invariable, uninterrupted fluoride level in the treated water.

CHAPTER 13 *Effects of Fluorides on the Distribution System*

The addition of fluorides to water is undertaken solely for the purpose of protecting children's teeth from dental decay. Since 1945, when this practice was begun, some interesting side effects and advantages to the water-plant operator have been experienced. None of these involves any additional or supplemental physiological benefits. They include means for detecting leaks and for establishing travel times in water-distribution systems.

LEAKS

It is a common experience in the operation of water utilities to find water standing or flowing on the ground near or over areas known to contain part of the distribution system. The source of such water may either be from a leak in the mains or be ground water. In order to determine definitely whether a leak has occurred without having to dig down to the main, a sample of this visible water is analyzed for fluorides. If the water in the community is fluoridated, then the source of this water is readily identified. If the sample contains no fluorides, considerable expense in digging to the main is avoided.

TRACING POSSIBILITIES

Many aspects of a distribution system cannot be readily determined by calculations involving consumption of water alone. Ordinarily it is impossible to determine accurately, or without considerable difficulty involving laborious calculations, information concerning a particular part of a system, such as water velocities, directions of flow or of no flow, position or presence of valves, and detention times in basins or mains. Much information of this type is relatively easily determined by the incorporation of a study related to the start of fluoridation in a community. By selecting critical points for sampling and knowing when fluoridation will begin, the first appearance of fluorides at these points can be readily established. This is done by taking frequent samples (about once every five minutes) at about the time the first appearance of fluoride has been predicted. By analyzing these samples on the spot and noting the time when the samples first contain about 0.2 to 0.3 ppm more fluoride than the raw water, the time of travel to the sampling point can be determined quite accurately. Such data can be used for improving the design of basins and the operation of pumping stations as well as determining the size and condition of mains (roughness coefficients) and their weaknesses from the standpoint of attaining maximum efficiencies and flows. It has been reported that this method has disclosed valves whose existence was not suspected.

SCALING

Incrustants in the form of calcium or magnesium fluorides or calcium or magnesium silicofluorides are formed occasionally in the pipelines, tanks, eductors, and pumps handling concentrated fluoride solutions prior to the point of fluoride application to the water supply. Scaling in the distribution system has never been known to have been caused by water treated by such fluoride compounds. The severity of scaling at the plant is a function of the type of fluoride compounds formed, the amount handled, the hardness of the water, and its velocity or degree of agitation at a given point.

Sodium fluoride causes the most voluminous and persistent scale. This scale is composed of either calcium or magnesium fluoride or a combination of both. Incrustants composed of calcium or magnesium silicofluorides (formed from the hardness constituents of the water and sodium silicofluoride) are the least troublesome because they are more soluble. Calcium and magnesium fluorides cannot form in solutions of sodium silicofluoride because of the lower pH values of these solutions. No scaling is experienced with hydrofluosilicic acid because this acid is not ordinarily diluted prior to feeding. The amount of scaling observed is related to the quantity of chemicals available to form the scale; that is, the concentration of the fluoride solutions and the amounts of calcium and magnesium in the water.

Scaling can also be caused directly by the deposition of calcium or magnesium fluorides on equipment because of the velocity of the water passing through. Centrifugal pumps and eductors together with the pipelines connected to them are frequently subjected to scaling, which disappears, however, when the installation is changed to a gravity system; this causes the water to flow more slowly. If such higher water velocities cannot be avoided, provision should be made for the easy removal of the piping for periodic cleaning.

In many cases, scaling is best eliminated by removing the original cause; i.e., either by reducing the amount of calcium or magnesium by softening the water prior to making the fluoride solutions or by sequestering these elements with hexametaphosphate solutions. Where relatively smaller quantities of make-up water are required (as, for instance, the water used in a sodium fluoride saturator tank), it may be more economical to soften the water in a small, salt-regenerated pressure softener. Where larger quantities of water are required (in the dissolving chamber of a dry feeder), polyphosphate solutions, made from Calgon, Calco 918, or Micromet, can be economically used. These require a small mechanical solution or pressure-pot feeder to ensure the most effective dosage of the water to the dissolving chamber. The amount to be fed (in the order of five to ten times the hardness) is best determined by experience.

CORROSION

The addition of the optimum quantity of fluorides to water supplies has never caused an increase in the corrosive properties of the water so treated. The concentrated fluoride solutions prior to feeding are corrosive, however, and materials in the feeding and piping equipment should be selected accordingly. Generally, the stainless steels and plastics furnished by the manufacturers of the equipment have been entirely satisfactory in overcoming this problem.

EFFECTS OF HOME WATER SOFTENERS ON FLUORIDE CONCENTRATIONS

Home water softeners contain an ion-exchange material which absorbs the calcium and magnesium ions from the water and releases sodium ions to the water in exchange. The sodium is replenished periodically (and the calcium and magnesium ions replaced and removed from the medium) by passing a concentrated brine (sodium chloride) solution through the medium bed. This process and chemical reaction do not ordinarily remove fluoride ions from the water.

Reports have been received, however, that in some few localities, the fluoride concentration in previously fluoridated water has been lowered after the water has passed through such softeners. Very little is known at present as to the cause of this. Inasmuch as the removal occurs only in certain well-defined areas, however, it appears that it may be caused by some contaminant in the water. From this observation it appears that the contaminant may be a fluoride-absorbing material which is filtered out by the softening medium. This forms a pad of defluoridation material on top of the softening material and removes the fluoride as the water passes through it. Whenever the nature of this contaminant is discovered, appropriate steps might be taken at the water plant to remove it from the water.

There have also been some reports that very occasionally the fluoride level is increased for short periods of time after passing through home water softeners. This phenomenon appears to be related to the method of regeneration at the plant where the softeners are serviced. The cause and remedy for this are also now being investigated.

TUBERCULATION IN MAINS

Depending on the composition of the water, tubercles (blisterlike protuberances on the inside of water mains) are sometimes formed on the interior surfaces of pipes over pitted areas; they consist of mounds of the products of corrosion. It has been shown that fluorides and other anions from the water can become concentrated in these tubercles by means not as yet completely understood. A possible explanation has been suggested:[1]

Dissolved oxygen has the greatest influence on the rate of corrosion of new mains. After a film of iron rust has been formed, however, other factors accelerating corrosion become more important. Pits with overlying tubercles start to form when the rust becomes so thick as

[1] John R. Baylis, Early Studies on the Corrosion of Iron Pipes, *Pure Water*, Department of Water and Sewers, Bureau of Water, City of Chicago, 9(9): 145–184 (1957).

to retard the diffusion of soluble iron compounds, such as ferrous sulfate, ferrous chloride, and perhaps ferrous fluoride, into the water from the pitted areas. These salts, being acidic, accelerate pit corrosion. At the same time the access of dissolved oxygen and additional anions from the water to the pitted area is blocked. This tends to make the corrosion by pitting self-limiting; the electrochemical reaction wherein hydrogen and anionic products are liberated must eventually cease because of lack of fuel, i.e., oxygen and anions from the water. It appears, then, that the anionic content of a tubercle reaches a maximum value and either remains at that point or decreases. In addition, the tubercles are very hard and quite insoluble in water.

An AWWA panel considering this matter in 1957 reached the following conclusions:

1. Fluoride ions may concentrate in tubercles in small amounts together with other negative ions, such as sulfates and chlorides.

2. The rate of concentration of these materials is so slow that the amount of fluoride removed cannot be measured by the usual test methods.

3. There is no way in which the fluorides concentrated in the tubercles can be remixed with the water so as to bring about a measurable increase in fluoride content.

CHAPTER 14 *Present Status of Fluoridation*

As shown in Fig. 14.1, the acceptance of fluoridation in the United States has increased steadily since the first installation in 1945. The growth has been particularly rapid since 1951, when the beneficial results were first announced. The most rapid growth occurred during the period 1950 to 1953 (Table 14.1). It was at this time that active, organized opposition to the measure began, with the result that during the past ten years a total of 96 water-supply systems elected to discontinue it, although 17 have reinstituted fluoridation after further consideration. At the end of 1961, 1,249 water-supply systems were furnishing fluoridated water to 2,197 communities where over 42 million persons lived. In addition, 7.3 million persons were using water containing naturally 0.7 ppm fluoride or more from public sources. This total of over 49 million persons using fluoridated water in amounts considered effective for caries control represents about 36 per cent of the people in the United States served by public water supplies.

Puerto Rico, with 93 per cent of its people served by fluoridated public water supplies, has the highest proportion of people receiving this benefit of any country in the world. Cities in many foreign countries have adopted fluoridation, but so far most of these (except in Latin America) are on an experimental basis.

A census of places fluoridating, including a list of names of such places in the United States, is published each year by the American Water Works Association.

The majority of the 128 communities that adopted fluoridation in 1960 were smaller cities and towns. Nevertheless, because a majority (93 per cent) of all communities have a population of less than

Present Status of Fluoridation

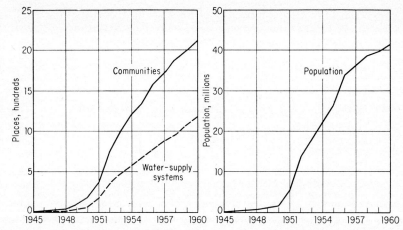

Fig. 14.1. Communities, water-supply systems, and population served with fluoridated water.

10,000, a large number of the smaller places have not as yet adopted fluoridation. For instance (Fig. 14.2), only 5.1 per cent of the places with fewer than 1,000 population are fluoridating. On the other hand, 31 per cent of all cities having a population greater than 10,000 have

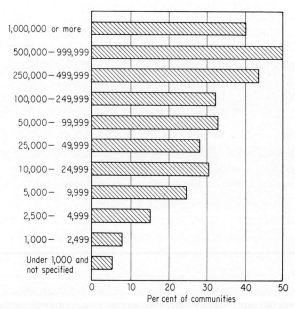

Fig. 14.2. Percentage of communities (by size) using controlled fluoridation. (USPHS.)

TABLE 14.1. ANNUAL CUMULATIVE FINDINGS ON THE INSTITUTION OF CONTROLLED FLUORIDATION, SHOWING NUMBER OF COMMUNITIES, WATER-SUPPLY SYSTEMS, AND POPULATION SERVED, 1945–1960

Year	Fluoridation status at end of each year		
	Number of communities	Number of water-supply systems	Population*
1945	6	3	231,920
1946	12	8	332,467
1947	16	11	458,748
1948	26	13	581,683
1949	49	29	1,062,779
1950	100	62	1,578,578
1951	368	171	5,079,321
1952	751	353	13,875,005
1953	1,007	482	17,666,339
1954	1,194	572	22,336,884
1955	1,347	672	26,278,820
1956	1,583	772	33,905,474
1957	1,717	879	36,215,208
1958	1,890	995	38,461,589
1959	1,990	1,081	39,628,377
1960	2,111	1,172	41,169,412
1961	2,197	1,249	42,201,115

* Includes adjustment for population growth during intercensal years 1951–1959 based on results of the 1960 census of population.

adopted it. Among the larger cities that have adopted fluoridation are Baltimore, Chicago, Cleveland, Denver, Indianapolis, Louisville, Miami, Milwaukee, Minneapolis, Norfolk, Philadelphia, Pittsburgh, Providence, Rochester, St. Louis, San Francisco, and Washington. Those which have not as yet provided fluoridated water include New York, New Orleans, Atlanta, San Diego, Seattle, and Boston.

An excellent summary of the operating experiences, methods, and costs at 20 of the larger cities was completed by New York University during 1957.[1] Of the vast array of information the following might be mentioned:

[1] Water Fluoridation Practices in Major Cities of the United States, prepared by New York University for the New York State Department of Health, December, 1958.

1. Fluoridation costs averaged $1.23 per million gal water treated (7.9 cents per person per year), which includes all items of expense connected with the feeding and control of fluoride compounds.

2. There is no evidence of fluorides accumulating in the mains or of any uncontrollable health hazards among waterworks operators who handle fluoride compounds.

3. All the cities studied were able to maintain the desired fluoride levels for long periods of time, and no instances of prolonged overdosages have been experienced since the projects were started.

Adoption of fluoridation is not confined to those communities which own their own water systems. At the end of 1961, 273 privately owned supplies were being fluoridated. An order by the various governing bodies continues to be the most widely accepted means for authorizing fluoridation. Resorting to referenda, however, appears to be increasing because of the controversy engendered by opponents.

CHAPTER 15 *Fluoridation of Individual Water Supplies*

So far in this book we have been concerned with the addition of fluorides to public water supplies. But there are two groups of people to whom this discussion has been of little use: those who are not served by a public water-supply system at all; and those whose waters naturally contain, not a deficiency, but an excessive concentration of fluorides. In the United States there are about 100 million persons now using water from approximately 20,500 public water-supply systems. The remaining population, about 80 million persons, is living in rural or suburban areas or in small towns where water supply is obtained from private, individual water sources, principally wells. It is entirely possible that water from a public supply will never become available to almost one-third of the people of this country. This proportion will probably not change significantly in the future because even though more public water supplies are constantly being established and expanded, additional families are also moving to rural or suburban areas where individual supplies are required.

Many alternative methods have been suggested for providing fluorides for this group. At the present time, the topical application of fluorides is widely used. By this means, a strong fluoride solution is applied directly to children's teeth. Compared with water fluoridation, it is not considered an effective public health measure in that it costs considerably more; the results obtained are in the order of a 40 per cent reduction in dental caries as compared with 60 to 65 per cent for fluoridated water; and it is difficult to provide treatment on a widespread basis. The other alternatives, described in Chap. 5, are even less effective or impracticable. They include the use of bottled

Fluoridation of Individual Water Supplies

fluoridated water, fluoride-containing pills, and fluoridated salt, baby food, vitamin pills, and milk.

The problems involved in the fluoridation of home water supplies are different from those encounterd in the fluoridation of public water supplies. The principal differences are that extremely small amounts of water are pumped and consumed; the types of installations vary widely; the operating personnel (if the home owners) are generally not trained in the techniques of water treatment; and the private water system is generally not maintained in a manner comparable to public systems.

However, a home water-supply system can be successfully and safely fluoridated if the following elements are incorporated:

1. A reliable and accurate system of feeding fluorides, with particular reference to the incorporation of means to prevent overdosing
2. Design of the entire installation to prevent tampering
3. Provision for monitoring the fluoride level in the treated water and replenishment of the fluoride solution

In order to obtain a reliable feeding system, the following criteria have been suggested:

1. The feeder should be capable of maintaining a fluoride level of 1.0 ppm with an error no greater than plus or minus 0.1 ppm (10 per cent) during all conditions of normal operation.
2. There should be a minimum possibility of leakage of water back through the feeder and into the fluoride-solution reservoir.
3. The siphon breaker provided must prevent the fluoride solution from entering the water lines unmetered during periods of negative pressure.
4. The feeder should not feed fluoride solution when no water is flowing despite the condition and operation of the well pump or the position of the well-pump motor starter. It is preferable to use a device for pacing the feeder proportionally to the water flow.
5. The equipment must be constructed of materials capable of withstanding the corrosive effects of the fluoride solutions.

It has been found that the following devices or combinations have been successfully used to meet these requirements:

1. The feeder (Fig. 9.8) is hydraulically operated, the driving energy being derived from the water pressure in the distribution system. The feeding diaphragm is actuated by another, larger diaphragm by means of the water pressure released by a solenoid-operated valve

on the water line. This solenoid is energized periodically either with a timer moving a microswitch or mercury switch or by means of an electric contactor on the shaft of a water meter (Fig. 9.6). With this type of feeder it is impossible to feed fluoride solutions when, for various reasons, the electric motor driving the well pump is running but no water is being pumped. This might happen, for instance, if the

Fig. 15.1. Small motor-driven chemical feeder. (*B-I-F Industries, Inc.*)

level of water in the well is too low for the pump to remove it or if the impeller or shaft of the pump is broken.

2. An electric-motor-operated feeder (Fig. 15.1), of which many different types are available, can be successfully controlled for this purpose by several means. For instance, an electric contactor in a water meter can be made to start a timer which will run during a predetermined and preset length of time for each contact and then reset itself. During the period the timer is running, it also makes the circuit which runs the feeder motor.

Fluoridation of Individual Water Supplies

3. The same type of feeder can be readily controlled by having it run continuously (as long as the well pump is running), but by varying the time during which it is alternately feeding fluoride solution and water (Fig. 9.13). The electric contactor of the water meter controls a three-way solenoid valve on the suction line to the feeder. When the contact is made, the valve opens to permit either fluoride solution or water to be pumped. When the valve is in the other position, the other liquid is pumped.

4. There are available various types of meter-operated feeders for very small installations (Fig. 15.2). These generally are of limited

Fig. 15.2. Meter-operated feeder. (*Economics Laboratories.*)

power, and the frequency of stroking is sometimes too small to provide a satisfactory fluoride level. However, where their use is possible, they provide the means for a very economical installation.

These combinations of apparatus, if properly maintained, have proved to be quite satisfactory from the point of view of accuracy. Some are obviously more complicated than others, and more expensive, but more experience with each type and with others which are constantly appearing will eventually reveal the best combinations for this type of service.

Despite the accurate performance of such devices, however, there must be available a service or system which will assure their continued satisfactory operation. Such a service should include the expert maintenance of the equipment, the preparation and replenishment of fluoride solution, and the constant surveillance of the resulting fluoride concentration in the treated water. These tasks are generally beyond the capabilities of the average home owner, and for this reason, he should purchase this service from an organization similar to those providing chlorination, water softening, or bottled water to individual home owners.

It is estimated that such a service would be profitable at a cost to the householder of between $3 and $6 per visit (once per month or every two months). This would result in a range in cost between $18 and $72 per year. The cost would be influenced by the number of customers on a route, the size of the area occupied by the customers, and the efficiency with which they were served. At the end of the period when continued fluoridation would confer relatively fewer benefits (the absence of children or the advancing age of the children), the installation could be removed, rehabilitated, and used again in another home.

CHAPTER 16 *The Practicality of*
Partial Defluoridation

The undesirable disfigurement associated with fluorosis and the resultant increased cost of dental care were known for several decades prior to 1931, when the cause was discovered. Soon afterward, it was shown that the severity of fluorosis was directly related to the fluoride concentration in the water used during the period of permanent-tooth calcification. Children continuously exposed to water containing about 5 ppm or more fluoride are invariably afflicted with mottled enamel of the permanent teeth; many of these children have gross calcification defects which weaken the enamel and cause eventual loss of teeth through attrition.

Even though methods for the removal of excessive fluorides from water were made available soon after the cause of fluorosis was discovered, little if any progress was made in the reduction of the incidence of fluorosis through these methods. This lack of acceptance of available measures for preventing fluorosis is probably due to the allegedly excessive costs of treatment plants, the costs of operating such plants, and the complexity of the operating procedures. Actually the experience we have so far gained in the design, construction, and operation of various processes for defluoridating water indicates that these allegations are no longer tenable.

At the present time there are approximately 4.1 million people living in 1,200 communities which are served by water sources containing fluorides in excess of the optimum. In 264 of these communities (in which 689,117 people live) the fluoride levels are at least three times the optimum. Despite the widespread use of such waters as sources of public supplies, only 11 plants designed specifically for fluoride removal are now operating.

At present, three defluoridation methods have proved practicable under varying conditions of raw-water quality and availability of treatment chemicals. These methods involve the use of activated alumina, bone char, or magnesium compounds. The first two methods employ insoluble, granular media which remove the fluorides as the water percolates through them. The media are periodically regenerated by chemical treatment when they become saturated with the fluoride removed from the water. In the third method, the fluorides are removed along with the magnesium which might be added in the form of a lime. Both the fluorides and the magnesiums are subsequently removed through the use of settling basins and then discarded.

Activated alumina is used at Bartlett, Texas, and at the Army plant at Camp Irwin, near Barstow, California. It is also used in several experimental home defluoridation units. This medium is available from several aluminum manufacturers in various mesh sizes of granules and degrees of purity. Activated alumina is commonly used as a desiccant, particularly in air-conditioning equipment. For our purpose, it is used in mesh sizes between 28 and 48. It is currently selling for $7\frac{1}{2}$ cents per pound and weighs, in place, 50 lb per cu ft. The Bartlett plant (Fig. 16.1) contains 500 cu ft of alumina in a standard circular steel filter tank 11 ft in diameter and $11\frac{1}{2}$ ft high. The plant is capable of treating 400 gpm of Bartlett well water (Edwards limestone formation) and was designed to reduce the fluoride from 8.0 ppm to an average of 1.0 ppm.

When between 400,000 and 500,000 gal water have been treated, the alumina becomes saturated with fluorides and must be regenerated. Regeneration of the activated alumina consists of various steps designed to remove, by backwashing, the accumulated solids that have been strained from the water; to remove (by means of a weak caustic solution) the fluorides that have been sorbed by the medium; and to neutralize the residual caustic with weak acid and water rinses. At present, the caustic solution is applied countercurrently, so that one operation combines the backwashing step (saving the treated water formerly used for this purpose) and the caustic application. After a rinse to reduce the alkalinity as much as practicable, a weak (0.05 normal) sulfuric acid rinse is applied until the alkalinity of the waste water is low enough to make the effluent usable. The total amount of raw water used for regeneration is 68,500 gal.

Recently a new well was drilled at Bartlett, reaching 2,670 ft to the Trinity sands. This water is similar to the older well water but contains only 3.0 ppm fluoride. One result of this change is that cycle lengths now extend to an average of 1.5 million gal. Because of the greater difficulty in removing fluorides at the lower concentrations, however, the capacity of the medium fell to 400 grains fluoride per

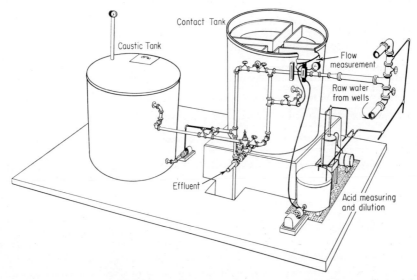

FIG. 16.1. Defluoridation plant at Bartlett, Tex. (*USPHS.*)

cu ft of medium from the 700-grain level when using the 8.0-ppm fluoride well.

The equipment for the Bartlett plant cost $11,360 in 1951, including installation. The alumina cost an additional $4,000. The building was furnished by the city. Operation started on March 11, 1952.

The plant at Bartlett is a gravity type in that the contact tank is open to the atmosphere. The plant at Britton, South Dakota, on the other hand, is a pressure type, in which the medium is confined in a heavy steel tank designed to withstand the pressure of water in the distribution system. The water at Britton is pumped from any of three wells through the defluoridation tank and directly into the distribution system. The tank contains 300 cu ft of bone char, which at present is used in the United States only at this plant. The fluorides are reduced from 6.7 ppm to an average of 1.5 ppm. The char is 30 to 50 mesh and is ordinarily used in sugar refineries for decolorizing the sugar syrup.

After treating about 450,000 gal, the bone char becomes saturated with fluoride and must be regenerated. This consists of backwashing to remove the accumulated sand and then pumping a 1 per cent solution of caustic soda through the medium. After the fluorides have been removed by the caustic, the excess caustic remaining in the bone char must be removed. This is done by first rinsing the bed with raw water until additional rinsing removes very little caustic; then the bed is treated with a weak carbon dioxide solution. This solution is prepared by passing carbon dioxide gas (from liquefied dry ice) through raw water as it passes a series of diffusers en route to the contact tank. When the pH of the waste water approaches that of the raw water, the plant is ready for another cycle. There are 6,500 gal treated water used for regeneration together with 27,000 gal raw water.

Construction of this plant was completed and operation was started November 20, 1948, with a synthetic hydroxy apatite for a medium. This material was abandoned during February, 1953, because of excessive (42 per cent per year) losses due to attrition. The cost of the equipment for the plant in 1947 was $12,245, including the bone char. Bone char of the type suitable for this purpose now sells for $4\frac{1}{2}$ cents per pound. A complete charge of 300 cu ft costs $540. Current chemical costs for regeneration amount to $53 per million gal.

No plants have yet been built specifically for fluoride removal using the magnesium process. There are, however, several softening plants in Ohio, Indiana, and Illinois that incidentally remove small amounts of fluoride along with the magnesium. This combination of functions, using the same equipment and treatment chemicals, would make the cost of defluoridation lower than any other process. On the other hand, the initial cost of building such plants is considerably higher, and chemical and sludge handling is more difficult.

The problems related to the operation, control, and maintenance of defluoridation plants are no more difficult than those encountered in conventional water-treatment plants. We have found that the employees of water systems previously involved only in well-pump operation and meter reading could become capable defluoridation-plant operators after adequate training. The equipment required for these defluoridation plants is a collection of standard water-treatment-plant items, and their cost and complexity are identical to those of softening or ion-exchange plants. Many plants have cited that operating costs for softening, iron removal, decolorization, clarification, or combinations of these processes are similar to costs of defluoridation.

The removal of excess fluorides from community water supplies to prevent dental disfigurement, loss of teeth, and increased cost of dental care is a sound public health procedure. The improved health of those using such water appears to require defluoridation in preference to many other common water-treatment processes.

APPENDIX *Health Objections*

A. MEDICAL

Practically every physical, mental, or moral illness to which human flesh is heir has been attributed by the opponents of fluoridation at one time or another to drinking fluoridated water. Those diseases, particularly heart trouble, nephritis, cancer, and mental disorders, the causes of which are unknown or poorly understood, are more frequently linked with fluoridation. A large amount of epidemiological work has been necessary to refute such completely unfounded statements. These studies have been published (references are a part of this appendix), and the results are available to everyone. The opponents of fluoridation, however, either have not referred to them or have claimed them to be in error or biased.

Data have been collected from a wide variety of places on the incidence of heart diseases, cancer, nephritis, and mental illnesses. There have been completed, in addition, several intense studies of the physical condition of people of all ages who have lived all their lives in towns supplied with naturally fluoridated water (at levels over eight times the optimum). There have been no noticeable untoward differences discovered except for dental fluorosis. None of the dozens of harmful effects and diseases attributed by those opposed to the use of fluoridated water have ever been confirmed, even in places having many times the optimum fluoride concentrations. The question has been summarized by Dr. R. A. Kehoe (Director of the Kettering Laboratory): "The question of the public safety of fluoridation is nonexistent from the viewpoint of medical science."

For those who are interested in referring to studies designed to demonstrate the absence of any relationship between various diseases and the use of fluoridated water, the following bibliography[1] is included.

Abortions

Wisconsin State Board of Health: Continuous-resident Data, 1951: Death Rates per 100,000 Population in Cities with Varying Concentrations of Fluorides in Public Water Supplies 1945–49.

This study indicates that in Wisconsin communities with fluoride in their water supplies ranging from 0.03 to 2.5 ppm, the presence of fluoride caused no consistent difference in the frequency of stillbirths, premature births, neonatal deaths, and infant or maternal deaths.

Acne

Epstein, Ervin: Effect of Fluorides in Acne Vulgaris, *Stanford Med. Bull.*, **9:** 243–244 (1951).

The rate of improvement among 40 persons with acne was the same among those who took prescribed fluoride tablets and those who did not.

Allergies

Hodge, H. C.: Fluoride Metabolism: Its Significance in Water Fluoridation, *J. Am. Dental Assoc.*, **52:** 307–314 (1956).
Report to the Mayor on Fluoridation for New York City, The Board of Health, The City of New York, 1955, pp. 28–29. (52 pp.)
Schlesinger, E. R., et al.: Newburgh-Kingston Caries-Fluorine Study. XIII. Pediatric Findings after Ten Years, *J. Am. Dental Assoc.*, **52:** 296–306 (1956).

There is no scientific evidence that edema or allergies are produced by drinking fluoridated water. Even in areas where the fluoride levels are eight times the optimum amount, allergies are no more prevalent than elsewhere.

Anemia

Newburgh-Kingston Caries-Fluorine Study: Final Report, *J. Am. Dental Assoc.*, **52:** 290–325 (1956).

Complete physical examinations of children who had been consuming fluoridated water for ten years at Newburgh, N.Y., revealed no evidence of anemia.

[1] A much more complete bibliography on these and other diseases and effects can be found in K. R. Ewell and K. A. Easlick, "Classification and Appraisal of Objections to Fluoridation," University of Michigan, June, 1957.

Arteriosclerosis—Hardening of the Arteries

Abraham, Albert: What the Dentist Should Know about Heart Disease, *J. New Jersey State Dental Soc.*, **27**: 32–34 (1956).

Some of the factors contributing to hardening of the arteries were known prior to the fluoridation of water supplies and are completely independent of fluoridation. Actually the occurrence of arteriosclerosis is a part of the process of aging. If the average life span increases, then the conditions that accompany aging will also tend to be more prevalent.

Arthritis

Steinberg, Charles L., et al.: No Relation Found between Arthritis and Fluoridation, *J. Am. Dental Assoc.*, **54**: 410–411 (1957).

> "These studies indicate that there is no relationship between various arthritic conditions and the musculoskeletal diseases. This study should dispel the fear that fluoridation of water, as recommended by health authorities, is a causation factor in arthritic conditions."

Asthma, Bronchitis, Tuberculosis, and Other Diseases of the Respiratory System

The medical literature does not contain a single instance that fluorides in drinking water are in any way related to these diseases.

Cataracts, Glaucoma, Conjunctivitis, Color Blindness, and Other Disorders of the Eye

Leone, N. C., et al.: Medical Aspects of Excessive Fluoride in a Water Supply, *Public Health Repts.* (U.S.), **69**: 925–936 (1954).

Examinations of the eyes for cataracts and lenticular opacities were made among people who had lived in a community with 8.0 ppm fluoride naturally occurring in the public water supply. There was no significant increase or prevalence of such disorders as compared with the rates in a town with only 0.4 ppm fluoride.

Schlesinger, E. R., et al.: Newburgh-Kingston Caries-Fluorine Study. XIII. Pediatric Findings after Ten Years, *J. Am. Dental Assoc.*, **52**: 296–306 (1956).

Over several years special ophthalmological examinations were made of a group of children consuming fluoridated water at Kingston, N.Y., for ten years. The findings of such examinations fell well within limits expected of any normal group of children in these age groups.

Diabetes

Water Fluoridation, Report of the Committee of the St. Louis Medical Society, Summary, pp. 338–360 in U.S. Congress. House Committee on Interstate and Foreign Commerce. Fluoridation of Water. Hearings before the Committee on Interstate and Foreign Commerce, House of Representatives, 83d Congress, 2d Sess., on H.R. 2341, a bill to protect the public health from the dangers of fluorination of water. May 25, 26, and 27, 1954, Government Printing Office, Washington, 1954. (vi + 491 pp.)

Wisconsin State Board of Health: Continuous-resident Data, 1951: Death Rates per 100,000 Population in Cities with Varying Concentrations of Fluorides in Public Water Supplies 1945–9.

No correlation was found in death rates for diabetes in Wisconsin cities with natural fluorides ranging between 0.5 and 2.5 ppm. In addition, no rise was discernible in diabetes prevalence after ten years of fluoridation at Sheboygan.

Digestive-tract Ailments Including Ulcers, Colitis, Nausea, Diarrhea, and Constipation

Leone, N. C., et al.: Review of the Bartlett-Cameron Survey: a Ten Year Fluoride Study, *J. Am. Dental Assoc.*, **50**: 277–281 (1955).

The Bartlett, Texas, study revealed that no unusual conditions of the digestive tract were found even though the water contained over eight times the optimum amount for decades.

Schlesinger, E. R., et al.: Newburgh-Kingston Caries-Fluorine Study. XIII. Pediatric Findings after Ten Years, *J. Am. Dental Assoc.*, **52**: 296–306 (1956).

Complete physical, laboratory, and radiographic examination of groups of children in the Newburgh-Kingston survey revealed no unusual conditions or abnormalities of the digestive tract.

Eye Troubles—see Cataracts

Goiter

May, Richard: Untersuchungen über den Fluorgehalt des Trinkwassers in bayerischen kropfgebieten verschiedener Endemiestarke, *Z. ges. exptl. Med.*, **107**: 650–651 (1940).

Murray, Margaret M., et al.: Thyroid Enlargement and Other Changes Related to the Mineral Content of Drinking Water, *Med. Research Council (Brit.), Mem.*, No. 18, 1948. (39 pp.)

Von Fellenberg, T. B.: Does Any Relation Exist between the Content of Fluorine in Water and Goiter? *Chem. Abstr.,* **33**: 2631 (1939).

These studies revealed that there is no correlation between the occurrence of endemic goiter and high concentrations of fluoride in the water and that there is no basis for the belief that low prevalence of goiter is associated with a low fluoride content of water and high rates are related to higher concentrations of fluorides.

Headaches

The medical literature does not reveal any instances of headaches caused by drinking fluoridated water, even up to eight times the optimum concentration.

Heart Disease

Arnold, F. A., Jr.: The Grand Rapids Fluoridation Study—Results Pertaining to the Eleventh Year of Fluoridation, *Am. J. Public Health,* **47**: 539–545 (1957).

Commission on Chronic Illness: Effects of Fluoridation of Community Water Supplies upon the Aged and Chronically Ill, Chicago, Commission on Chronic Illness, Mar. 17, 1954. (3 pp. processed)

Hagan, T. L., Pasternack, Morton, and Scholz, Grace C.: Waterborne Fluorides and Mortality, *Public Health Repts. (U.S.),* **69**: 450–454 (1954).

Hodge, H. C., and Smith, F. A.: Some Public Health Aspects of Water Fluoridation, pp. 79–109 in J. H. Shaw (ed.), "Fluoridation as a Public Health Measure," American Association for the Advancement of Science, Washington, 1954. (v + 232 pp.)

Schlesinger, E. R., et al.: Newburgh-Kingston Caries-Fluorine Study. XIII. Pediatric Findings after Ten Years, *J. Am. Dental Assoc.,* **52**: 296–306 (1956).

These reports relate to comparisons of morbidity and mortality rates for heart disease between cities with and without fluorides. No significant differences could be found.

Hepatitis and Cirrhosis and Degeneration of the Liver

Hagan, T. L., Pasternack, Morton, and Scholz, Grace C.: Waterborne Fluorides and Mortality, *Public Health Repts. (U.S.),* **69**: 450–454 (1954).

There is no evidence in the literature that fluoridated water has any connection with these disorders.

Intracranial Lesions

Hagan, T. L., Pasternack, Morton, and Scholz, Grace C.: Waterborne Fluorides and Mortality, *Public Health Repts.* (*U.S.*), **69:** 450–454 (1954).

Knutson, J. W., Statement of, pp. 273–276 in U.S. Congress. House Committee on Interstate and Foreign Commerce. Fluoridation of Water. Hearings before the Committee on Interstate and Foreign Commerce, House of Representatives, 83d Congress, 2d Sess., on H.R. 2341, a bill to protect the public health from the dangers of fluorination of water. May 25, 26, and 27, 1954, Government Printing Office, Washington, 1954. (vi + 491 pp.)

A study of the incidence of such lesions in 64 cities with fluoride levels below 0.25 ppm and above 0.7 ppm showed that no differences were discernible. In addition, no differences in death rates from such causes were found in studies made in Michigan, Wisconsin, Illinois, and Texas.

Kidney Diseases—see Nephritis

Liver Diseases—see Hepatitis

Multiple Sclerosis

It has never been shown that multiple-sclerosis incidence has been influenced by the consumption of fluoridated water.

Nephritis, Nephrosis, Uremia, and Other Diseases of the Kidneys

Hagan, T. L., Pasternack, Morton, and Scholz, Grace C.: Waterborne Fluorides and Mortality, *Public Health Repts.* (*U.S.*), **69:** 450–454 (1954).

Heyroth, F. F.: Toxicological Evidence for the Safety of the Fluoridation of Public Water Supplies, *Am. J. Public Health,* **42:** 1568–1575 (1952).

Illinois Department of Public Health, Bureau of Statistics: Mortality in Fluoride and Non-fluoride Areas. Springfield, Illinois, Department of Public Health, Health Statistical Bull., Special Release, No. 20, Apr. 1, 1952. (8 pp. processed)

Wisconsin State Board of Health: Continuous-resident Data, 1951: Death Rates per 100,000 Population in Cities with Varying Concentrations of Fluorides in Public Water Supplies 1945–9.

The departments of health of Illinois and Wisconsin have not detected from their vital statistics any differences in nephritis rates from cities with varying amounts of fluorides in the public water supplies. It has been concluded that "No evidence exists that waterborne fluorides has been a cause of Nephritis."

The Newburgh-Kingston, N.Y., study revealed no evidence of any adverse effects of fluoridated water on the kidneys.

Osteosclerosis

Hodge, H. C., and Smith, F. A.: Some Public Health Aspects of Water Fluoridation, pp. 79–109 in J. H. Shaw (ed.), "Fluoridation as a Public Health Measure," American Association for the Advancement of Science, Washington, 1954. (v + 232 pp.)

Schlesinger, E. R., Statement of, pp. 31–33 in "Our Children's Teeth: A Digest of Expert Opinion Based on Studies of the Use of Fluorides in Public Water Supplies," Committee to Protect Our Children's Teeth, Inc., New York, March 6, 1957. (vi + 104 pp.)

Steinberg, C. L., et al.: Comparison of Rheumatoid (Ankylosing) Spondylitis and Crippling Fluorosis, *Ann. Rheumatic Diseases*, 14: 378–384 (1955).

Specific studies on this condition have shown that there is no connection between the incidence of osteosclerosis and the use of fluoridated water. In the Newburgh-Kingston examinations, there were no detectable differences in the bone densities of the children in the two cities—one fluoridating, the other without fluorides. Hodge and Smith stated, "Between the amount of fluorine that will produce osteosclerosis in humans and the amount obtained by drinking fluoridated water (1 ppm) there is a safety factor of 8- to 20-fold."

Poliomyelitis

No single incident has ever been reported which established a relationship between using fluoridated water and poliomyelitis.

Psychoses, Neuroses, Neuritis, Neuralgia, and Nervous Disorders

All of the long-term medical studies on the possible effects of fluorides reported no relationship between these mental disorders and the presence of fluorides in drinking water.

Respiratory System—see Asthma

Rickets

Schlesinger, E. R., et al.: Newburgh-Kingston Caries-Fluorine Study. XIII. Pediatric Findings after Ten Years, *J. Am. Dental Assoc.*, 52: 296–306 (1956).

The pediatric findings at Newburgh, N.Y., revealed no significant bone changes in the people consuming the fluoridated water as compared with those who did not.

Sterility

There has been no reported instance of sterility produced by consuming fluoridated water, including waters naturally fluoridated to levels many times the optimum.

Tuberculosis—see Asthma

Varicose Veins

There has never been any evidence produced to show that fluoridated water causes varicose veins.

B. ALLEGED DENTAL EFFECTS

It has been alleged that dental caries is either unaffected or actually increased by the use of fluoridated water. The following ten-year studies prove that this allegation is exactly the opposite from what actually occurs:

Arnold, F. A., Jr., et al.: Effect of Fluoridated Public Water Supplies on Dental Caries Prevalence, *Public Health Repts. (U.S.)*, **71:** 652–658 (1956).
Brown, H. K., et al.: The Brantford-Sarnia-Stratford Fluoridation Caries Study—1955 Report, *J. Can. Dental Assoc.*, **22:** 207–216 (1956).
Council on Dental Health, Fluoridation Program Begins for Greater Cleveland Residents, *Dental Health Highlights*, **12:** 25–28 (1956).
Hagan, T. L., Pasternack, Morton, and Scholz, Grace C.: Waterborne Fluorides and Mortality, *Public Health Repts. (U.S.)*, **69:** 450–454 (1954).
Hill, T. J.: "A Textbook of Oral Pathology," 3d ed., Lea & Febiger, Philadelphia, 1945. (407 pp.)

In addition, there are numerous other reports which relate to other cities which have examined children before and after fluoridation and which report similar reductions in the rate of carious teeth. These include:

Hill, I. N., Blayney, J. R., and Wolfe, W.: Evanston Dental Caries Study. XVI. Reduction in Dental Caries Attack Rates in Children Six to Eight Years Old, *J. Am. Dental Assoc.*, **53:** 327–333 (1956).

Two More Studies Show Fluoridation Effective, *J. Am. Dental Assoc.*, **52:** 767 (1956).

Dental Deterioration

Kronfeld, Rudolf: "Histopathology of the Teeth and Their Surrounding Structures," 3d ed., Lea & Febiger, Philadelphia, 1949. (514 pp.)

The claim that the dentine might deteriorate because of fluoridated water might have been based on the supposition that fluorides could cause a withdrawal of calcium from the teeth. It has been established that withdrawal of calcium from the teeth is not possible, either because of fluoride consumption or for other reasons.

Malocclusion

Pelton, W. J., and Elsasser, W. A.: Studies of Dentofacial Morphology. III. The Role of Dental Caries in the Etiology of Malocclusion, *J. Am. Dental Assoc.*, **46:** 648–657 (1953).

Zimmerman, E. R., Leone, N. C., and Arnold, F. A., Jr.: Oral Aspects of Excessive Fluorides in a Water Supply, *J. Am. Dental Assoc.*, **50:** 272–277 (1955).

A survey of malocclusion conducted in fluoride and nonfluoride areas concluded that "while the dental-caries experience differed, the dentofacial index (a measure of malocclusion) was essentially the same in both communities." Even in areas with excessive fluorides (8.0 ppm at Bartlett, Texas) malocclusion was no more in evidence than normally experienced.

Mottled Enamel

Dean, H. T.: The Investigation of Physiological Effects by the Epidemiological Method, pp. 23–31 in F. R. Moulton (ed.), "Fluorine and Dental Health," American Association for the Advancement of Science, Washington, 1942. (101 pp.)

Report of Ad Hoc Committee on Fluoridation of Water Supplies, National Academy of Science, National Research Council, Washington, 1952. (8 pp.; National Research Council Publication 214)

A mottling of the tooth enamel has been known since 1931 to be caused by an excessive intake of fluorides consumed during periods of calcification of the permanent teeth. As the fluoride level in water increases, the severity of the mottling increases until at the level of about 5.0 ppm, almost all children have moderate or severe mottling. At 1.0 ppm, fewer than 10 per cent of the children have the least detectable evidence of such mottling, a degree that can be detected accurately only by an experienced dentist and that is not in any way aesthetically undesirable. In hot, dry climates where children drink more water than in cooler places, the optimum level should range downward to about 0.7 ppm rather than 1.0 ppm.

Periodontal Disease

Russell, A. L., and Elvove, Elias: Domestic Water and Dental Caries. VII. A Study of the Fluoride–Dental-caries Relationship in an Adult Population, *Public Health Repts. (U.S.)*, **66:** 1389–1401 (1951).

Russell, A. L., Statement of, pp. 24–26 in "Our Children's Teeth: A Digest of Expert Opinion Based on Studies of the Use of Fluorides in Public Water Supplies," Committee to Protect Our Children's Teeth, Inc., New York, March 6, 1957. (vi + 104 pp.)

Fluoridated water does not appear to contribute any effect one way or another toward the development or suppression of periodontal disease. Its incidence at Colorado Springs (2.6 ppm fluoride), for instance, was found to be similar to that of Baltimore, where the fluoride level at the time of the examination was practically zero.

C. VETERINARIAL

Goldfish and Tropical Fish

Kleinhenz, J. E.: Discussion of Fluoridation Experiences from a Manager's View Point, a Panel Discussion, pp. 603–608 in Task Group E5-10, "Committee on Fluoridation Materials and Methods," Proceedings of the Annual Conference at Kansas City, Mo., May 7, 1952, *J. Am. Water Works Assoc.*, **44:** 595–616 (1952).

Goldfish living in an aquarium filled with fluoridated water from the Indianapolis water supply were unharmed. This is corroborated by the more than 2,200 communities now fluoridating their supplies, where there was not a single instance of harm to fish or other animals when using fluoridated water.

Rabbits, Mice, Rats, Chickens, Dogs, Swine, Sheep, Cattle, Goats, and Monkeys

National Research Council, Committee on Dental Health: The Problem of Providing Optimum Intake for Prevention of Dental Caries, a Report of the Committee on Dental Health of the Food and Nutrition Board, Division of Biology and Agriculture, National Research Council, Washington, 1953. (15 pp.; National Research Council Publication 294)

Schmidt, H. J., Newell, G. W., and Rand, W. E.: The Controlled Feeding of Fluorine, as Sodium Fluoride, to Dairy Cattle, *Am. J. Vet. Research*, **15**: 232–239 (1954).

Rabbits grow normally on rations containing 200 ppm fluoride; swine at the 300 ppm fluoride level. Maximum dosages which may be tolerated by various animals, in milligrams per kilogram of body weight, are: rats, 10 to 20; guinea pigs, 12 to 20; dairy cattle, 1 to 3; chickens, 35 to 70; swine, 5 to 12.

The Conference of Public Health Veterinarians (Nov. 14, 1956) stated that the consumption of water fluoridated at the recommended levels is not harmful to pets or *any* other animals.

Index

Accuracy, of feeders, 114–118
   effects on dental results, 114
   gravimetric feeders, 103, 113
   overall, 116–117
   solution feeders, 99
   tests for, 115–116
   volumetric feeders, 103
  of fluoride analysis, standards, 159
  of visual instruments, 165
Adoption procedures, 26–31
Adults, excluded from benefits, legal interpretations, 34, 36, 40
  possible benefits from fluoridated water, 20–21
  confirmation of, 51
Age distribution in Grand Rapids study, 17
Alternate methods for administering fluorides, 204–205
  bottled water, 46–47
  bread and salt, 47
  chewing gum, mouth wash, tooth paste, 45
  milk, 47
  tablets (pills), 46
  topical applications, 45
Alum solutions, for dissolving fluorspar, 95
  effect of, on added fluorides, 156
  on fluoride analysis, 171
Alumina, activated, for defluoridation, 210–211
Aluminum industry, production of fluoride compounds by, 40
  role in discovery of cause of fluorosis, 4
American Dental Association, 30, 38
Ammonium silicofluoride, calculations with chlorine, 86
  characteristics, 82, 86
  costs, 82, 86, 88
  manufacturers, 87
  other uses, 86
  preparation, 86

Ammonium silicofluoride, shipping containers, 87
  unsaturated solutions, 93
  use (census), 81
Analysis of fluorides, 158–187
  automatic, 182–186
  basic theory, 160–162
  Megregian-Maier, 165, 170, 172–174
  role in maintenance of fluoride levels, 41
  Scott-Sanchis, 170–172
  sources of error, 176–177
  SPADNS, 169–170, 174–176
Apatite, source of fluoride compounds, 5, 60–61
  source of natural fluorides in water, 48, 60
  (*See also* Phosphate rock)
Application of fluorides, points of, 155–157
Armstrong, W. L., 14
Arnold, F. A., fluorosis observations, 19, 24
  fluorosis at optimum fluoride levels, 55
  Grand Rapids study, 17
Ast, David B., 11, 13
Aurora, Ill., natural fluoride results, 18, 21
Automatic control, of chemical feeders, 126, 133–136
  electric, 147–148
  pneumatic, 148
  of fluoride levels, 182–187
   colorimetric system, 182–184
   conductivity system, 184, 186–187

Baby teeth (*see* Deciduous teeth)
Bag loader for feeder hoppers, 152, 191
Baking industry use of fluoridated water, 43
Baltimore, Md., survey, industrial use of fluoridated water, 43

Bauxite, Ark., fluorosis prevalence, 2
Baylis, J. R., 198
Bellack, E., 181
Bone char in defluoridating water, 211–212
Bottled water, fluoridated, 46
Brantford, Ont., first city suggested for fluoridation, 11
　control cities in conjunction with study, 16
　plan of study, 12–13

Calcium fluoride (see Fluorspar)
Campaigns for fluoridation, 27, 29
Canning industry use of fluoridated water, 44
Caries, experience, Dean's studies of, 11–13
　reduction, tables, 16, 22, 23
　(See also def rates; DMF rates)
Carlos, James P., 24
Characteristics, of communities in fluoridation campaigns, 27, 29
　of fluoride compounds, 60–90
Charlotte, N.C., ice cracking with fluoridated water, 43
　seasonal changes in fluoride levels, 56
Chewing gum, fluoride-containing, 45
Chiaie's disease, description, 1
Chicago fluoridation suit, 38
Chlorine, and fluoride in water treatment, 40
　interference with fluoride determination, 157, 171
　simultaneous feeding of, 74, 208
Chrietzberg, J. E., 114
Churchill, H. V., discovery of cause of fluorosis, 34
City engineer, role in fluoridation programs, 29
Color measurement systems, 161
Color standards for fluoride analysis, 163–165
　costs, 165
　disadvantages, 164, 165
　Hellige tester, 165
　Taylor comparator, 164
Colorado Springs, Colo., fluorides in water supply, 49
　role in fluorosis research, 2
Colorimetric systems of fluoride analysis, 166, 169, 183
Compounds, fluoride, 4, 60–90
　blue-tinted, 193
　chemical objections to, 39–40

Compounds, fluoride, computation for determining quantities, 123, 158–159
　detention times required, 119
　distribution in world, 10, 60
　selection, 87–90
Computation, for adjustment of feeders, 124
　for chemical quantity, 123
　of dissolving chamber capacities, 119
　of feeder accuracies, 116–117
　of feeder capacities, 92, 124
　for feeder selections, 123
　of fluoride level, 158–159
　of optimum levels based on temperatures, 58–59
Concentration of fluoride, correct, benefits from, 51
　laboratory procedures for determining, 158–187
　maintenance, constant, at Grand Rapids and Newburgh, 117–118
　natural, 49–50
　optimum, determination of, 48–49
　variation according to temperature, 56–59
Constitution of the United States and fluoridation, 33–34, 36, 38
Control, automatic (see Automatic control)
Conveying system for fluoride compounds, 190
　types and manufacturers of, 151
Conway, B. J., 38
Corrosiveness, relative, of fluoride compounds, 89
　of water after fluoridation, 42
Cost, of chemicals, 82, 88
　of fluoridation, average, 203
　of repairing each DMF tooth, 50
Courts, and fluoridation, 32–38
　U.S. Supreme, 32, 38
　state, 33–35, 38
Cox, G. J., early suggestions for fluoridation studies, 11
　research on animals with fluorides, 4
Criteria in fluoridation studies, 15
Cryolite, characteristics and commercial sources, 87
　occurrence and use of, 49
　source of natural fluorides in water, 48, 60

Dall tubes, 138–141
　costs of, 141
　recorders for, 141

Index

Dean, H. Treadley, caries observations, 11, 16
   conclusions based in studies, 19–20
   fluorosis, in Georgia, 56–57
   index and investigations, 7–10
   studies in 21 cities, 10–11, 14, 52
Decantation type solution feeders, 96, 101
Decayed, missing, and filled tooth index (*see* DMF rates)
Deciduous teeth, def rates, 15
   effect of fluorides on, 21, 24
   at Grand Rapids, Mich., 23
def rates, 15
   reduced, with fluoridated water, 21, 24
   suggested mechanism for, 24
Dental health status in U.S., 26–27
Dental society role in fluoridation promotion, 27
Department of Defense fluoridation policy, 31
Design of fluoridation installations, 50
Diaphragm solution feeders, 96–97
Dilution, of fluoride water to optimum levels, 49
   of water samples for interference reduction, 177
Dissolving chambers, 118–121
   capacities and detention times, 118–119
   design, 119
   necessity for, 118
   removal of fluoride solutions from, 153
Distillation of water samples, 177–181
   continuous, apparatus, 185
   critical features of equipment, 178–179
   direct, 180, 181
   Willard and Winter method, 177
Distribution system, effects of fluoridated water on, 195–199
DMF rates, 15
   effect on, of feeder accuracy, 114
   from 57 studies, 50
   observations, at Brantford, Ont., and Newburgh, N.Y., 19
   at Grand Rapids, 17–23
   reduction per unit increase in fluoride levels, 51

Eager, J. M., 1
Eductors, operating requirements for, 153
Effects, physical, of fluoridation, 25, 215–225
Equipment, safety, 150–151, 188–192
Errors, in fluoride analysis, 176–177
   in fluoride feeding (*see* Accuracy)
Exposure, fluoride, in water plants, 188

Feeders, accessories, 148–154
   accuracies, 114–118
   auxiliary equipment, 125–154
   control, 125–148
   gravimetric, dry, 92, 103, 112–114
   range (capacities), 92
   screw-type, 108–109
   solution, 92–103
      for home systems, 205–207
   types of, 91–124
   vibratory, 106–108
   volumetric, 104–111
Flooding of feeders, prevention of, 193
Flow splitters for dissolving chamber effluents, 154
Fluoride(s), administration of, 45–47, 204–205
   analysis of, 158–187
   compounds (*see* Compounds)
   contributing to water naturally, 4, 7
   excessive, in water, 10, 209–210
   ingestion, prenatal effect of, 24
   ion(s), chemistry of, 5
      description of, 5
      differences between, 5, 39–40
      interfering, 171, 177–180
   minerals, 5, 60
   occurrence in foreign countries, 10
   recovery of, from phosphate rock preparation in, 77
   selection of, for fluoridation, 87–90
Fluorosis, description and degrees of, 2
   in 44 U.S. cities, 8–10
   index, for obtaining optimum fluoride levels, 55
   resulting from fluoridated water, 24–25
Fluorspar, formation from silicofluoride solutions, 196
   mineral, 61–71
   characteristics, 61, 69, 82, 88
   costs, 63, 69, 82
   deposits, 62
   distribution, 49, 61
   historical use, 61–62
   manufacturers, 70, 71
   preparation for use, 62–63

Fluorspar, mineral, safety in handling, 189
  use in water treatment, 63–69
  source of fluoride ion in chemical manufacture, 4
  source of natural fluorides in water, 48, 60
Fluosilicic acid (hydrofluosilicic acid), characteristics of, 77–78, 82
  comparable effectiveness, 77–78, 82
  corrosion effects, 78–79
  costs, 80, 82, 88
  dilution precautions, 79–80
  equipment cost savings, 80
  manufacturers, 81
  materials suitable for handling, 79
  preparation, 78
  safety when handling, 78–79, 189, 191
  shipping, 80–81
  specifications, 79
  storage tanks, 88
  use (census), 81
Food, Drug and Cosmetic Act and fluorides, 37
Food and Drug Administration, policy on fluoridated water use, 37
  ruling on fluorides in water, 43
Foods, fluorides in, 52–53
Frequency of sampling for fluoride analysis, 159

Gallagan, D. J., 58
Gentile tubes, 138–139
Grand Rapids, Mich., consistency of fluoride levels, 117
  control city used, 16
  plan of study, 13, 15
  results of study, 16–24
    results compared with other cities, 16, 19
Gravimetric feeders, 92, 112–114
  accuracy, 103, 113
  capacities, 112
  costs, 112
  manufacturers, 114
  principles of operation, 103, 112–113
Growth of fluoridation, numbers of communities and populations, 201
Gypsum, formation with alum and fluorspar, 63, 66
  removal from fluorspar solutions, 67
  separation from fluorspar, 68, 95

Haggaman, W. H., 77
Hannan, Frank, 3
Hardness of water, in fluoride losses, 89
  in scaling, 197
  upper limit for economical softening, 149
  (*See also* Softening equipment)
Hazards, of fluoride compounds, 90
  to water-plant operators, 150–151
Herschel, Clemens, 139
Hexametaphosphates, in lessening fluoride losses, 89
  in scale prevention, 149
Hill, W. L., 75
History, of fluoridation, 1–13
  of fluorosis observations, 2–4
Home water supplies, defluoridated, 210
  fluoridated, 204–218
    costs, 218
    need for, 204
Hopper, feeder, bag loader, 191
  filling, 190
Hutton, W. L., 11, 13
Hydraulically actuated feeders, 134, 136
Hydrofluoric acid, 71, 78, 80, 191
  method of feeding, 95
Hydrofluosilicic acid (*see* Fluosilicic acid)

Ice cracking with fluoridated water, 43
Impulse-duration systems for feeder control, 133, 136
Incrustants, formation of, in fluoride compounds, 89
Ingestion of fluorides, constancy, 51
  variations with age, 52, 54
Interferences, in fluoride determinations, 177
  list of values, 171
  methods of control, 177–181
Ion (*see* Fluoride, ion)
Ionization, 5

Jordan, H. E., 42

Knowlton, K. F., 186
Knutsen, J. W., 19

Laboratories, role in fluoridation, 50, 158–160

Index

Laboratory procedures (*see* Analysis of fluorides)
Lantz, E. M., 4
Leak detection with fluoridated water, 195
Litigation on fluoridation, 32–38
  references to, 38
Louisiana Supreme Court and fluoridation, 34

McClure, F. J., 53, 85
McKay, F. S., 2, 3, 6
Magnesium silicofluoride, characteristics, costs, and present uses, 82, 87, 88
  source and shipping, 87
  unsaturated solutions, 93
Magnetic-flow meters, 141–142
Maintenance of fluoride levels, 114
  examples of, 117, 118
  necessity for, 114
Masks, approved for fluoride dusts, 191
  filters for, 151
Mass medication and fluoridation, 36
Media for providing fluorides, 25, 44–47
Medical examinations at Newburgh, N.Y., 24
Medical objections to fluoridation, 215, 222
Medical practice and fluoridation, 32
Medical significance of fluoridation, 25
Medical society role in fluoridation, 27
Medication, forced, and fluoridation, 34
Meters, compound, 130
  displacement, 127
  indicating, for dissolving chambers, 121
  magnetic-flow, 141–142
  oscillating-piston, 127–128
  propeller, 130–132
  velocity, turbine and current, 128–130
Microswitch for pacing, 132–134, 136
Milk, as carrier for fluorides, 47
  natural fluoride content, 53–54
  milk teeth (*see* Deciduous teeth)
Mixing equipment, for dissolving chambers, 120, 188
  in fluoride analysis, 172–173, 175, 182
  for fluorspar dissolvers, 64, 68
  for preparing solutions from dry fluoride compounds, 149
Mottled enamel (*see* Fluorosis)

Murdock, J. H., 38
Muskegon, Mich., 16–18, 22–23

Natural fluorides, versus manufactured fluorides, 39–40
  presence in water, 4, 7, 48
  source of, 48–49, 60
Need for fluoridation, 26–27
Nessler tubes, 162–163
Newburgh, N.Y., constancy of fluoride levels, 26
  plan of study, 11, 13, 26
Nozzles, insert, 138–139
  restriction, in pneumatic signal transmission, 144–145

Oakley, Idaho, fluorosis prevalence, 2
Objections to fluoridation, chemical, 39–41
  economic, 45–47
  engineering, 42
  health, 215–225
    dental, 32, 222–224
    medical, 32, 34, 36, 216–222
    veterinarian, 224–225
  industrial, 42–44
  legal, 32–38
  religious, 34–35
Opponents of fluoridation, reasons for, 29
  answers to objections of, 29–30
Optimum fluoride levels (*see* Concentration of fluoride)
Orifice plates, 138–139
  pressure change through, 140
  recorders for, 141
Oroville, court case at, 35

Pacing of fluoride feeders, 126, 136
  means for, 133–137
Pallets for compound storage, 154
Parts per million (ppm), 4
Personal rights and fluoridation, 34–35
Petrey, A. W., 4
Phillips, R. S., 56
Phosphate rock, composition, 75
  fluoride content, 75
  occurrence and origin, 49, 75
  preparation for use, 74–77
  reserves, 75
  source of natural fluoride in waters, 48
Phosphates, effect on fluoride analysis, 171

Photometers, 165–169
  costs, 167
  filter, 167
  operation, 168
  precautions, 176–177
  spectrophotometers, 166
  theory, 166–167
Piston-type solution feeders, 96, 98–99
Placental transfer of fluorides, 24
Plants, fluoride uptake, 54
Police power and fluoridation, 32–33
Populations, using fluoride compounds, 81
  using water, with controlled fluorides, 200–202
    with excessive fluorides, 50, 209
    with natural fluorides, 50, 200
Potassium fluoride, costs, 82
  preparation, 71
  use with chlorine in water treatment, 74
Ppm, explanation of, 4
Precipitate formation when diluting fluosilicic acid, 79, 90
Pressures allowable for solution feeders, 97–98
Primary devices, 126–142
  displacement meters, 126–137
  pressure-differential devices, 137–141
  magnetic-flow meters, 141, 142
Privately owned water plants, fluoridating, 203
Procedures, suggested, for adopting fluoridation, 29
Public health and fluoridation, 32, 34
Public opinion and fluoridation, 26
Puerto Rico, status of fluoridation in, 200
Pulsating flow from solution feeders, 98

Recorders for pressure-differential devices, 141
Reduction, in def rates, 24
  in DMF rates, 14–24
  in fluorosis, 209
Referenda, 26, 203
Religious freedom and fluoridation, 32, 34
Removal of fluorides from water, 209, 213
  with coagulants, 156
  communities requiring, 49
  costs, 211–212
  need for, 49, 209

Roper, Elmo, fluoridation survey, 26
Rotating cup solution feeders, 96, 100

Safety, to consumer, 192–194
  equipment for handling fluorides, 150–151, 188–192
  of fluoridation, 37
  of water-plant operators, 188–192
Sampling, water, frequency, 159–160
Saturator tank for sodium fluoride, 89
  computing quantity of solution used from, 149
  description and operation, 93
Scaling, in distribution systems, 196–197
  in fluoride solution lines, 89
  from silicofluorides, 89, 196
Schlesinger, E. R., 25
Scientific organizations approving fluoridation, 27–30
Screw-type volumetric feeders, 105–108
Seafoods, fluorides in, 53
  at Tristan da Cunha, 52
Seasonal fluoride levels, 59
Selection, of complete fluoridation installation, 91
  of fluoride compounds, 87–90
Sewage treatment, effect of fluorides, 44
Shipping containers, disposal, 192
Signal transmission, 142–147
  electrical, 145–146
  pneumatic, 144–145
Silicon tetrafluoride in manufacture of silicofluorides, 77–78
Siphoning, prevention, 155
Sludge removal, from saturator tanks, 150
  from solution lines, 157
Slurry feeding of fluoride compounds, 95
Smith, M. C., and Smith, H. V., 4
Sodium fluoride, characteristics, 72–73, 82
  costs, 82, 88
  manufacturers, 74
  other uses, 73–74
  populations using, 81
  preparation, 71
  safety in handling, 189
  scaling characteristics, 196
  shipping containers, 73
  specifications, 73
  tinting, 73

Index

Sodium fluoride, unsaturated solutions, 93
   germicidal effects, 157
   use in saturators, 72
Sodium silicofluoride, characteristics, 82, 83
   comparable effectiveness, 72, 85
   corrosion characteristics, 85
   costs, 82, 88
   feeding characteristics, 84, 85
   manufacturers and distributors, 85–86
   other uses, 84
   populations using, 81
   preparation, 84
   scaling experiences, 196
   shipping and containers, 85
   solubility and slurries, 83, 88
   specifications, 84
Softening equipment, for home systems, effects on fluoride levels, 197–198
   to prevent fluoride losses, 89
   to prevent scaling, 197
   for saturators, drawing, 94
   with solution feeders, 149
Sognnaes, R. F., 53
Solubility, fluoride compounds, comparison, 82, 88
   sodium silicofluoride, with temperatures, 83
Solution feeders, 92–103
   accuracy, 99
      tests, 115–116
   computations for adjustment, 124
   costs, 82, 100–101, 103
   description of types, 95–96, 100–103
   for home water systems, 205–207
   manufacturers, 99–100, 102
   for producing solutions of fluorspar, 66–68
   for use, with fluosilicic acid, 103
      with slurries, 95
      with solutions of dry compounds, 93
Space for feeders, limitations, 88
Stadt, Zachary, 56
Stallsmith, W. P., 38
"Standard Methods for the Examination of Water and Wastewater," 161
Standards, accuracy, for fluoride analysis, 159
   of feeder accuracies, 115, 193
   of fluoride compounds (see particular compound)
   safety, for fluorides in air, 188
Status of fluoridation (census), 200–203

Storage bins, factors in determining size, 122–123
Striffler, D. F., 50
Suction box, float-controlled, 155
Supplementing existing fluorides in water, justification, 50
Supreme Court, of Louisiana, 34, 38
   of United States, 32, 38
   of Washington, 33–34

Tablets (pills), fluoride-containing, 46
Tea, fluoride content, 52–53
Temperature, calculation of effect, 58–59
   and optimum fluoride levels, 55, 58
   raising, of water for dissolving chambers, 121
   seasonal fluoride levels based on, 59
   and sodium silicofluoride solubility, 83
Tinting of fluoride compounds, 193
Tooth paste, fluoride-containing, 45
Topical applications of fluorides, 45, 204
Toxicity, of chlorine, 40–41
   of fluorides, 40
   hazards to operators, 42
Tracing possibilities with fluoridated water, 195–196
Transmitted (absorbed) light, 161, 166, 168
Tuberculation in water mains, 42, 198

Vacuum breakers, 67, 193
Valves, air-meter operated, 135, 187
   solenoid, 133–134, 148
Variations in fluoride levels in water, 193
Velu, H., 4
Venturi, G. B., 137
Venturi tubes, 138–141
   cost, 141
   laying lengths, 141
   recorders, 141
Vibrators for dry compound storage hoppers, 152
Vibratory volumetric chemical feeders, 106–108
Visual color-measurement systems, 162–164
Volumetric feeders, 92, 103–111
   accuracy, 103
   capacities, 105–106, 108, 110–111
   costs, 106, 108, 110–111

Volumetric feeders, description of types, 105–111
  manufacturers, 105–106, 108, 110–111

Water, consumption, among children, 25
  variations with age, 54, 57
  hot, for feeder dissolving chambers, 121
  quality, cause of variations, 60

Water, supplies, individual, fluoridation of, 204–208
Water-plant superintendent, in fluoridation, 19
  in fluorosis investigations, 3
Wells, fluoridation of water from, 155
Williams, D. B., 19

Zipkin, I., 72
Zirconium-alizarin lake in fluoride analysis, 169–170